The Man Who Filmed Nessie

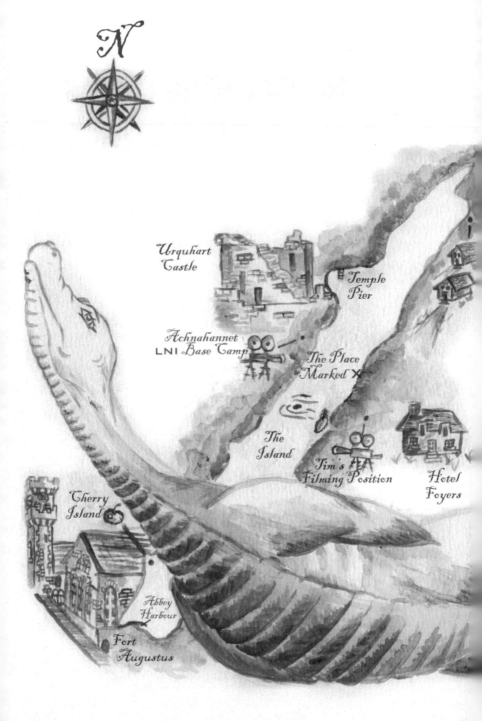

The Man Who Filmed Nessie

Tim Dinsdale and the Enigma of Loch Ness

ANGUS DINSDALE

hancock house

ISBN 978-0-88839-727-0
Copyright © 2013 Angus Dinsdale

Library and Archives Canada Cataloguing in Publication

Dinsdale, Angus, author
The man who filmed Nessie : Tim Dinsdale and the enigma of
Loch Ness / Angus Dinsdale.

Includes bibliographical references and index.
Issued in print and electronic formats.
ISBN 978-0-88839-727-0 (pbk.).— ISBN 978-0-88839-726-3 (html)

1. Dinsdale, Tim. 2. Aeronautical engineers—Great Britain—
Biography. 3. Loch Ness monster. I. Title.

QL89.2.L6D55 2013   001.944   C2013-902409-3
                                            C2013-902410-7

Editor: Theresa Laviolette
Production and cover design: Ingrid Luters

COVER PHOTOGRAPHS:
Front cover, top: Tim Dinsdale, photo by Peter Byrne.
Front cover, bottom: Tim Dinsdale, self-portrait photo.
Back cover: Tim Dinsdale aboard *Water Horse*, photo by Ivor Newby.

*We acknowledge the financial support of the Government of Canada through the
Canada Book Fund for our publishing activities.*

Printed in South Korea—PACOM

hancock
house

Published simultaneously in Canada and the United States by
HANCOCK HOUSE PUBLISHERS LTD.
19313 Zero Avenue, Surrey, BC Canada V3S 9R9
1431 Harrison Avenue, Blaine, WA, USA 98230-5005
Tel: (604) 538-1114 Fax: (604) 538-2262

www.hancockhouse.com | sales@hancockhouse.com

Contents

DEDICATION

*For my mother, Wendy, whose resolve made possible
my father's amazing quest.*

Foreword

In the 1950s when Tim and I married we were like most post-war young couples, looking forward to an exciting new life and having a family. We began our married life in Canada and enjoyed the time we had there, made even more fun with the births of our first two children, Simon and Alexandra. After five eventful years we returned to the UK and settled into the delightful countryside of Berkshire. While Tim continued his career in the aeronautical industry, I looked after the family, to which we had added another daughter, Dawn. With three children under five, life was seldom dull.

Then, in 1960, something happened which changed our lives. After researching the subject of a living unknown animal in Loch Ness, and later while visiting the loch, Tim took a film of something unidentified swimming there. I clearly remember the evening he telephoned from Loch Ness with the news of what he had seen and filmed. His natural exuberance and excitement was catching but perhaps my response was a little muted, being more interested, at that point, in when he would be starting his journey home. I was seven months pregnant and beginning to feel the need for a little support.

The media expressed considerable interest, and a couple of months later Tim was asked to appear and show the film on a BBC program called *Panorama*. Although the program was not airing until the evening, because it was broadcast live he had to spend the day in London while the family waited in anticipation at home for the show to begin. It was exciting for all of us, and not just because Tim was on television for the first time—the newest member of the family also decided this would be a good time to join in the fun. So, unbeknownst to my husband, immediately after watching the program I got myself to hospital and Angus made his own appearance in the wee hours of the morning.

One could say that day featured the entrance of two very different creatures into our lives—one small and cute, the other anything but—and one put our family life on a very different path.

Tim's first experience at Loch Ness had left him in no doubt that there was something extraordinary waiting to be discovered. My opinion was that if he had filmed the animal after only six days at the loch then surely it would not be all that difficult to film it again, only this time closer and clearer, so how hard could it be to get this over and done with? Obviously Tim thought the same, but how wrong we both were!

We could not know then the effect the pursuit of Nessie would have on our lives in the years to come. Tim's career in aeronautics came to an end, although a new one opened up as an author and lecturer; however, the financial rewards were much reduced, something to consider when one is responsible for a family. I gave my support by taking care of the family while he was away, and later by embarking on a career of my own, which helped to ease Tim's burden of responsibility. His drive and passion to find out the truth about the mystery of Loch Ness never failed and he pursued this quest until his untimely death.

This book will, I hope, be of interest to many different people, for it is not only about one man's search for the Loch Ness monster; it shows a family man who gave his children a great interest in and a curiosity of life, and he always made sure they had lots of fun while doing it. For many years our summer holidays were different from those of our friends. Not for us sunny holidays abroad, yet I recall no complaints when school broke up and once again we were all heading north to engage in various experiments and taking turns watching the loch, hoping we could help "capture" Nessie. Tim gave much pleasure to a generation of young people from his many talks to schools on that and other subjects. The book also uncovers his family history and his unusual childhood—surely there are not many people who can claim to have been pirated in the South China Seas.

As Angus weaves his way through the years I believe he has captured the essence of his family, and most of all his father, Tim Dinsdale, a man who took his work at Loch Ness seriously but who always maintained the fun side of our lives.

It is not often that a parent is privileged to see their life and the lives of the family they have brought up, through the eyes and memories of an adult child. Angus has certainly revived many of those for me. I do hope you too will enjoy reading about an ordinary family with extraordinary interests!

— Wendy Dinsdale
Reading, UK
January 2013

Preface

It's June 13, 1960, and Wendy Dinsdale lies, in labour, in the maternity room at Battle Hospital in Reading, England. At the end of her bed the doctor picks up her hospital notes and starts reading. When he gets to her name he stops, looks up and says, "That's interesting. I've just been watching someone called Tim Dinsdale on *Panorama* talking about a film he took of the Loch Ness Monster." Wendy, in considerable discomfort and with slight annoyance replies, "Yes, that's my husband."

Two months prior to the BBC *Panorama* programme, Tim, fuelled with a curiosity born from an article he'd read in *Everybody's* magazine a year earlier, had set off from his home in Reading to explore the infamous waters of Loch Ness. After twelve months of researching the subject, Tim's hope was to get a clearer understanding of the fabled monster by taking a firsthand look at the loch and perhaps meeting and interviewing some of the folks who claimed to have seen the "beastie"— or "Nessie" as the locals affectionately call her. He had prepared himself for the expedition by equipping his car as a mobile filming platform, removing the passenger seat and replacing it with a tripod and a 16mm Bolex ciné camera. Tim would be ready if indeed Nessie decided to show during his visit—and show she did!

On his first expedition to Loch Ness, Tim took an incredible film sequence recording something large moving across the surface waters of the loch. Whatever it was he filmed that day can clearly be seen moving powerfully, travelling in a zigzag pattern toward the north shore, eventually submerging and disappearing out of sight.

And so right there my lifelong relationship with Nessie began. I arrived in the early hours of June 14 to a much-changed family. My

11

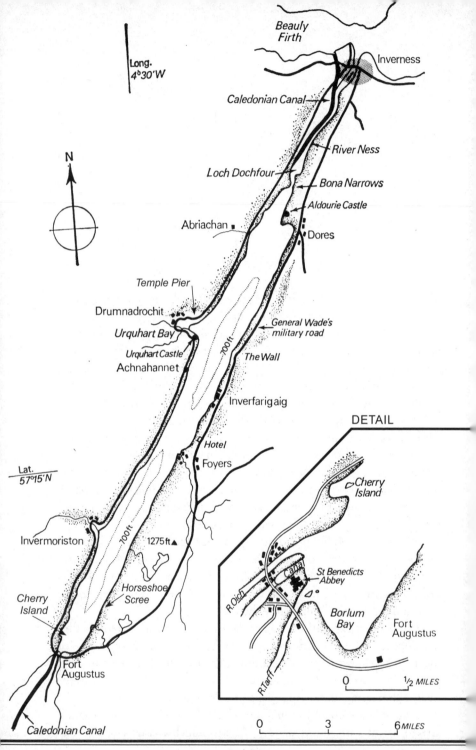

Beauly
Firth

Inverness

Long.
4°30'W

Caledonian Canal

River Ness

Loch Dochfour

Bona Narrows

Aldourie Castle

N

Abriachan

Dores

Temple Pier

Drumnadrochit

General Wade's
military road

Urquhart Bay

Urquhart Castle
Achnahannet

The Wall

700ft

Inverfarigaig

Hotel

DETAIL

Foyers

Cherry
Island

Lat.
57°15'N

St Benedicts
Abbey

R. Oich

Canal

Invermoriston

1275ft▲

700ft

Horseshoe
Scree

Borlum
Bay

Fort
Augustus

R. Tarff

Cherry
Island

Fort
Augustus

0   ½ MILES

Caledonian Canal

0   3   6 MILES

Loch Ness

12

father, by profession an aeronautical engineer, had just been on national television (which in those days was quite something) showing his recently taken film of a legendary monster living, swimming and breathing in the depths of the largest body of water in the British Isles. Let me put that into perspective for you: today's equivalent would be appearing on *Oprah*, saying not only had you met extraterrestrials but you also had the film to prove it, and then showing yourself shaking hands with ET. Some people would take it at face value and believe it; others would be sceptical and pay no attention to it; and then there'd be those who would shout hoax, immediately dismissing it outright. Well it's been no different with my father's film and the whole subject surrounding the existence of Nessie—or more scientifically, the existence of a large aquatic animal unknown to science living in a landlocked lake in a heavily populated western country. It just doesn't sound right and generally the scientific community likes to dismiss it, ignore it, or laugh at it. Only on very rare occasions does someone dare to explore it a little further.

Over the years since my father's film was first aired on the BBC, there have been countless monster-hunting expeditions to the loch with many excited and willing individuals and groups traveling north in the hope of finding out what is lurking—if anything—in those inky black waters of Loch Ness.

CHAPTER 1

Not Your Typical Summer Holiday

I grew up going on monster-hunting expeditions. Many of my school summer holidays were spent in a variety of caravans, guest houses or boats at the loch-side, and one thing I remember is that it seemed to be nearly always raining. Scotland is not the place to go if you're looking for good weather, even in the summer. But then sunshine and sand were not our holiday goal—monster hunting was!

Nevertheless, my brother and sisters and I were in our element and had tremendous fun spending our summers doing something totally unique. Our dad was off all day monster hunting, doing all sorts of exciting experiments, meeting interesting people from all parts of the world, being interviewed for French TV, helping an American group by towing their side scan sonar from his boat, *Water Horse*. He even assisted with the laying of hormone baits in the hope of recording an amorous Nessie via hydrophone. But best of all, we often got to go along. And so there I was, a cheeky kid with a killer grin, monster hunting, driving *Water Horse* across the loch while my sister Dawn was on watch with a camera at the ready (someone was always on watch) and Dad was inside the cabin fixing some gadget or other, mumbling and blaming the Loch Ness gremlins when it wouldn't work—for the umpteenth time.

For us kids it was an endless round of pure adventure meeting characters from all around the globe. There was Wing Commander Ken Wallis, a Second World War bomber pilot who'd designed and built a tiny one-man autogyro contraption. It was an amazing piece of machinery powered by a Rolls Royce Continental 0-200-B 100 hp engine which to a young boy didn't look all that much bigger than that of a standard garden lawnmower. Ken would tow it behind his Austin Mini, find a convenient farmer's field, start the motor by hand flipping the prop, climb aboard, and off he'd go racing across the grass until, quite quickly, he'd be airborne, buzzing around the skies skimming the tops of trees. I think the idea was to do some low-level flying across the loch in the hope of sneaking up and catching an aerial glimpse of the ever-elusive Nessie. I remember the autogyro causing quite a stir with locals and tourists alike, as it hadn't been all that long since Ken and his flying machine had featured in the 1967 James Bond movie, *You Only Live Twice*.

There was also Ivor Newby, a pleasant gentleman and an accomplished scuba diver. Ivor became known as "Ivor the Diver," not just because of his skills under the water, which were considerable—anyone who dives in Loch Ness has my respect—but also due to the fact that he made a habit of falling into the loch. Slipping off his boat was almost a common occurrence, but walking off the end of a pier gained the biggest cheer and stood him quite a few pints at the local tavern.

Many who came to the loch in search of the monster were inspired by Tim's 1960 film. Verification of the film's authenticity by the Royal Air Force's world-renowned Joint Air Reconnaissance Intelligence Centre (JARIC)[1] in 1966 helped to heighten interest worldwide—not to mention the boost it gave to the film's credibility and Tim's reputation. JARIC reported the object filmed was neither a surface boat nor a submarine and so was probably animate. In 1972, Tim also sent the film to NASA's Jet Propulsion Laboratories in Pasadena with similar results.

In the intervening fifty years since Tim took the now legendary sixty-five-second sequence, all sorts of people—professional bodies, universities, TV networks, and so on—have tried their luck at finding that one piece of evidence strong enough to convince not only the world's scientific establishment but also a sceptical public that Nessie is alive and well, enjoying herself in Scotland's very own Jurassic Park. Many have arrived at the loch full of enthusiasm, convinced their particular experiment or method of research will be the one to solve this timeless legend, only to have their hopes dashed, leaving empty handed—and, more often than not, rather damp. Results from these innumerable experiments, scientific studies and thousands of hours of shore watching

1 Now known as the Defence Geospatial Intelligence Fusion Centre (DGIFC) as part of the Joint Forces Intelligence Group (JFIG)

can only be described as intermittent; many were sketchy—meaning "nothing." But, on very odd occasions, there have been stunning and quite literally take-your-breath-away moments, leaving you scratching your head in wonderment: What exactly is down there lurking in those murky depths?

What is remarkable is that after years of effort and the massive advances in technology, and some very intriguing results (such as Rines' 1972 underwater flipper picture backed up with a sonar trace of a thirty-foot-long target showing projections or humps), Tim's film remains arguably the most convincing piece of evidence of a large aquatic animal inhabiting the loch.

In 1993, six years after Tim's untimely death, a TV company in the UK doing a story on Loch Ness took a copy of the film and did its own investigation. One of the technical operators noticed something and switched the film over to view it in negative. He then laid a number of the same frames over one another to reveal a shadow underwater. This shadow is unmistakeable in its shape, clearly showing a large diamond-shaped fin, a body, and what appears to be a tail, just under the surface. For someone who's literally had a lifetime of the monster, and viewed my father's film from all angles, it was startling to see. I was—and still am—amazed by that image.

So there we were, an unassuming family living in a modest 1950s-style semi-detached house in a quite ordinary cul-de-sac just on the outskirts of Reading (sounds like the opening to a Harry Potter story) suddenly thrust into the limelight due to an extraordinary subject getting worldwide attention—creating, of course, quite a stir on our street. Because of the unusual nature of monster hunting, some folks thought it was cool and were curious, wanting to know more, while others politely smiled and said nothing. Then there was the third group, the "take the mickey"[2] brigade. You would expect your mates to "have a go" about hunting monsters, but it didn't stop there; adults were also quite okay with having a dig about it, even to the point where one of my schoolteachers belittled the subject in front of the rest of my class. My reaction was of course to defend both my father and the subject, but play-

---

2 To tease or ridicule.

ground taunts can be harsh. However, in my school friends' defence, it was tough for other kids to understand what we did for our summer holidays. Remember, this was long before mass-market package tour holidays to Spain, Portugal, and Greece of which the British are so fond nowadays; most families at that time would take their summer breaks somewhere in the UK—a fortnight in Devon, a week at Butlins or, if you were really lucky, a trip to one of the Channel Islands. Few of my school friends, if any, went to Scotland—and certainly *none* went monster hunting. So while other kids talked about what they got up to during the six-week summer break, and spoke about getting a fancy flavoured ice cream in Swanage or building sand castles on Bournemouth beach, I would regale them with stories of being on watch, spending hours keeping my eyes peeled just in case Nessie (a prehistoric monster!) popped her head up while I was there, ready and waiting with a camera to capture the moment. My audience was rather sceptical and somewhat cynical. I guess it all seemed a little dramatic, and so un-Reading like.

My father suggested the mocking was likely due to a lack of understanding and knowledge on the subject, so to combat the naysayers he offered to give a talk to the entire school, kids and teachers et al, about "Nessie hunting." I thought that was a splendid idea, and on the following day made an appointment to see the headmistress to explain why it would be in the school's interest to learn more about the Loch Ness monster. The headmistress agreed and a date was set for my father to come and address the school. I was ten years old!

I was never really sure whether it was my dad talking about a cool subject like monster hunting, or the fact that we got the afternoon out of classes to listen to him that made my school life different. Either way my kudos jumped tenfold with my peers, and my teachers were so enthralled with Nessie they gave us a project on the subject. Of course I was the main source of information and, as you can imagine, went from being the butt of everyone's jokes to the most popular kid in the class—for a few weeks anyway.

My first expedition to Loch Ness had been in the summer of 1967; I'd just turned seven and, being the youngest of the Dinsdale clan and full of energy, was pretty much kept on a tight leash. We stayed in the hotel at Foyers where my father had resided during his first visit to the loch in 1960. Sitting high on the south shore, the old hotel commanded a stunning panoramic view of Foyers Bay and the loch in both a westerly and easterly direction. Dad was camping on a small island at the mouth of the Foyers River. This was a favourite spot of his, as it was the area where he'd seen and filmed the beast seven years earlier; it also gave a great deal of privacy as no one knew he was there and the island was only accessible by boat. Simon, my eldest sibling, spent some days and nights with my father on the island; it helped to have a second person to break up the boredom and the loneliness of the endless hours of the monster hunter's vigil.

Mainly for safety reasons, Dad carried a pack of maritime distress flares. Being mostly alone on the island he needed to have a way of contacting the outside world if anything went awry. Firing a flare was also a secret message to the folks at the Loch Ness Investigation Bureau (the LNIB was a group of dedicated individuals who were also involved with studying and hunting Nessie, of which I'll explain more as the pages turn). Their headquarters was situated on the north shore about four or so miles in an easterly direction. If they saw a flare it would indicate that Tim had either filmed or sighted the monster.

One morning while on early watch—5:30 a.m. to be exact—Dad decided to show Simon how to use the flare gun, and test fired one of the new batch of flares. He was confident it was early enough in the morning not to be seen by any of the LNIB crew. A huge ball of fire filled the sky, illuminating the surface of Loch Ness and much of the surrounding mountains before slowly falling back to earth. However, unbeknown to Tim, the LNIB had a team out doing early morning boat drifting and had seen something they couldn't explain and so assumed it was the monster. They grew very excited when, just moments after their sighting, a flare shot into the sky from the vicinity of the island. Knowing full well that was the spot of Tim's camp, they rushed back to Fort Augustus at the westerly most point of the loch and called the LNIB HQ with the news. Then, in turn, the LNIB contacted the hotel at Foyers asking to talk with Wendy, my mother, to try and confirm the report. Of course Wendy knew nothing of this, but was thrilled that perhaps at last Tim had snared the monster. Amidst great excitement we all packed into the car and hurried down the road to Foyers Point where we rushed to the water's edge, calling and shouting over to the island trying to get Simon's and Dad's attention. On seeing us they quickly launched the small boat and paddled over to find out what the hullabaloo was about. There ensued a comical moment with both my parents talking completely at cross purposes, Mum thinking Dad was playing coy pretending nothing had happened, and Dad bewildered by everyone's behaviour. Once the story was straightened out and we all realized nothing had been seen nor—worse—filmed, disappointment set in. Mum took me along to a nearby phone box where she made the call to the LNIB folks letting them know it was a false alarm. I remember they also needed convincing that nothing of interest had occurred. Leaving the phone box, both Mum and I were deflated and more than a bit saddened that the excitement of catching the monster had not been realized. As Dad would say many times in the coming years, "For a fleeting moment we thought we had Nessie by the tail and yet she slipped from our grasp."

And that was it: I was part of a monster-hunting family. It sounds bizarre, but that was the way of things for many years to come. Dad would continue to go on expeditions at least once a year, and sometimes the family would go along and others times not. He would lecture at universities and schools across the UK, Europe and in the USA. A steady stream of strange and interesting folks would come to our home to meet him and discuss their own expeditions, experiments and ideas to gain evidence that Nessie existed. Some of these people were invited to stay for dinner; others were politely given an hour of his time and then shown the door. Back then, just as now, the subject had a habit of attracting some less than honest folk looking for an angle. Tim occasionally was guilty of being a little too trusting; however the less scrupulous met their match when confronted by Wendy!

Whenever Dad returned from a trip, whether it be near or far, long or short, he always had a dramatic story to tell: a tremendous storm blowing up on the loch in the middle of the night that nearly put himself and *Water Horse* on the rocks; interviewing a witness who actually had to fend off the monster with an oar, breaking it in the process; or simply a description of the next piece of "very important equipment." Each subject was dealt an equal level of importance, seriousness, and, of course, humour. Ours was a household full of adventure.

Chapter 2

Slow Boats

It has always intrigued me how someone with an engineering background—an aeronautical engineer, to be exact—with a solid career in a cutting-edge industry (indeed, aeronautics was a most exciting business to be involved with during the fifties and sixties: fighter jets were breaking the sound barrier and heading toward Mach 2, the Harrier Jump Jet was on the drawing board, as was Concorde, and Rolls Royce was leading the development of the first generation of commercial jet engines; it was the computer industry of the time) would turn their back on a successful career and everything they had worked toward to go and chase monsters.

Tim wasn't a fool, a dreamer, nor taken for flights of fancy; he had responsibilities, a young family to provide for, and of course he had to meet all the financial trappings that go along with raising four children. He certainly wasn't independently wealthy; and yet he was pulled strongly enough in a direction of pursuing the truth and passion of belief that eventually he would give up the security that regular nine-to-five employment brings to concentrate his efforts full time on a quest some would say was as foolhardy as searching for the Holy Grail. Others eyed his courage to step out of the mainstream as enviously as if they were watching someone who'd just won the lottery.

To understand how and why such a transition took place I didn't have to dig too far into his colourful colonial-style childhood and up-bringing to appreciate that Tim was an adventurer from the get-go. Tim was born in Aberystwyth on the windswept coast of Wales in 1924, while his parents, Dorys and Felix, were on a year's leave from their post in China. Tim was their second child; his sister Felicity had been born in Hong Kong a year earlier. Felix, a shipping agent for Butterfield & Swire had himself been born in Yokohama in Japan and had only visited Britain once before, in 1918, to join the Royal Flying Corps at the end of the First World War.

And so when Tim decided to enter the world, Felix and Dorys were duly making the rounds of the country, visiting relatives, many of whom Felix had only ever heard about and as yet never met. Dorys' sister, Win, lived in a lovely cottage right on the beach in a tiny village called Borth, just up the coast from Aberystwyth. It was Borth where they were staying at the time of the birth. Felix, thinking ahead to the days when he would be bringing his family home on a more permanent basis, fell in love with Wales, and Borth in particular. While walking one day he spied a rather grand looking house perched high on the cliff overlooking the bay and the miles of sandy beach that add to the charm of the small seaside town. He proclaimed to Dorys that was to be their home when they finally returned from duty in the Far East. The house was called Sea Haven and had been built by a retired sea captain. After the captain's death, the house was split into two flats and years later Dorys, then a widow, purchased the upper level and lived there enjoying the spectacular view to a grand old age. We had some wonderful family holidays visiting Borth and staying in the old captain's house where there were many stories of the captain's ghost being seen on the staircase. I certainly remember being spooked when entering from the back door and having to negotiate the dark stairwell which had that musky smell of older houses. My sister Dawn and I would rush up to see who could get to the top fastest. It wasn't really a race; we were both just trying to get past the "haunted" part as quickly as our little legs would carry us.

As Tim approached his first birthday he was bundled up and, along with his sister and numerous travelling trunks, made his first trip to the Far East on the proverbial slow boat to China. An eight-week passage ensued through the Mediterranean, the Suez Canal, and the Red Sea, on to the Gulf of Aden and the Arabian Sea before crossing the Indian Ocean and finally entering the South China Sea and docking at journey's end in Hong Kong. Stops were made at various ports along the way, and with the changing landscapes came the changing of cultures from Western to Middle Eastern and Indian to Asian. It was about as far removed from modern travel as you could imagine when today a ten-and-a-half-hour flight gets you from London to Beijing—nonstop.

From Hong Kong, the expanding Dinsdale family was posted to Antung where Tim's younger brother Peter was born. These, however, were troubled times in China. A socialist uprising in 1927 prompted Felix to pack Dorys and the children off to Shanghai and then on to Hong Kong for their safety. The country was on the brink of revolution; millions of peasant farmers were revolting against the landlords, and in the cities the trade unions were taking control of the factories with industry coming to a virtual standstill. The attempted revolution, a precursor for the Chinese Communist Party takeover in later years, was suppressed by the Chinese National Army who turned against their own in bloody retribution to regain control for the imperial state.

Once the troubles had subsided and the country returned to some sort of stability—although China continued to be plagued with a guerrilla-style civil war for the next decade—Dorys returned to Antung where the family settled into the typical expat lifestyle of the day with the three children being educated at home by a governess. By all accounts it was an idyllic early childhood. The children played happily together within the safety of the walled garden. They would rarely venture outside, and certainly never alone, as lurking was an ever-present danger: European children were a rich prize, prime targets for kidnappers to snatch and hold for a tidy ransom. In spite of this threat, Felix would periodically insist on taking the family into the countryside for picnics. After all, it was the British thing to do! The countryside was known for harbouring bandits, so Felix's one concession to his family's safety was to always carry a revolver in the glove compartment of his Model T Ford.

An understanding of the dangers that surrounded their daily lives served Tim well during his first real adventure. With the Dinsdale children getting bigger, their education was to be continued at the China Inland Missionary School in Chefoo, which was some distance from their home in Antung and so required a journey by ship to attend. Returning for the spring term of 1935, Tim, Felicity and Peter, along with seventy other (mostly expat) children waved farewell to their parents who were standing on the dock at Shanghai as the SS *Tungchow* departed for its 500-mile trip up the coast to Chefoo. The ship was laden with all manner of goods and, along with the children returning to school, the decks were crammed with Chinese passengers claiming a spot for the journey.

Shortly before 6:00 p.m. on the first day of sailing, a drama started to unfold, which would cause ripples around the world. Shots rang out; it was the signal for a group of stowaway pirates to start fighting with the handful of ship's guards for control.

A gang of over twenty desperadoes had concealed themselves by boarding the ship under the guise of ordinary Chinese passengers and, once at a safe distance from port, they struck in a well-rehearsed attack. Splitting into three distinct groups, all with separate responsibilities, they fought for control of the ship. One group took care of the passengers by herding them into the saloon, while another wrestled for supremacy on the bridge, and a third undertook a pitched battle with the small contingent of the ship's guards, killing one and wounding several others before finally taking command.

Securing both the bridge and the radio room, the villains set the *Tungchow* on a new course, sailing toward the infamous pirate's lair of Bias Bay. Located just sixty miles north of Hong Kong, Bias Bay was where the South China Sea pirates nested and plied their dastardly trade for almost 100 years. Pirated ships were brought back to the bay and stripped of anything of value before being released. Rich passengers, however, whether European or Chinese, were always held on to and sold for ransom.

Once the gunfire had ceased and all the commotion had calmed down, the pirates set about searching the ship for both weapons and loot. The children, along with their group leader and some of the other passengers, were kept in the saloon where a revolver-wheeling buccaneer ordered them to stay and behave.

The pirates had learned the *Tungchow* would be carrying a large quantity of money in the form of bank notes, and there was indeed $250,000 in one-yuan notes on board. However, what the pirates didn't know was the notes were all incomplete; the cash was in transit from London, England to the Bank of China where the final part of the printing process would take place to make them legal tender. Without this procedure the notes were worthless.

On the second day, the pirates tried to camouflage the ship to avoid detection. They painted over the design on the funnel, adding two white rings, and changed the ship's name. This caused some amusement among the school children as the pirates misspelled the replacement name. The intention was to call her "Toya Maru"; however, due to their lack of language skills they painted the name backwards with it coming out as "uraM ayoT." The attempts to conceal the ship's identity were unnecessary as they sailed on un-harassed, sighting no other vessels for the next two days.

When the *Tungchow* had been out of radio contact for over twenty-four hours, and calls to her went unanswered, the alarm was raised. The likelihood was the ship had been pirated; the large number of European children on board would make a rich prize for any bandit willing to risk taking on the colonial powers.

With the news of the missing ship the British Royal Navy swept into action, deploying the aircraft carrier HMS *Hermes* to search for the *Tungchow* and its precious cargo. The ship had disappeared, and for all the authorities and the parents knew, the children were being held for ransom.

Day two passed without incident. The pirates were more relaxed and started to joke a little with the passengers, while the children, due to being cooped up all day, were getting restless. When a request went out for food the pirates rummaged through the supplies and found barrels of oranges. They took great delight at pelting the schoolboys with them, who in turned enjoyed the game of both dodging the incoming projectiles and trying to be good wicket keepers by catching as many as they could.

The light mood of the second day changed on the third to a more serious and almost sinister one as the ship sailed into Bias Bay. The pirates started firing their weapons at passing junks trying to capture one to ferry them, and their booty, ashore. This proved not such an easy task since manoeuvring the much larger and slower *Tungchow* into a position to give chase gave the junks time enough to turn tail and run. One gutsy junk captain was bold enough to return fire using his own antiquated cannon. After more than an hour of trying to commandeer a junk, and using much ammunition in the process, the pirates finally caught one. The captured junk came alongside to load aboard anything of value—including the $250,000 of worthless notes. At this point the children were robbed of any cash they had. One of the young girls later wrote, "us girls gave $109, the pirates in return gave us big oranges—mine cost $7..."

Once everything was loaded, six of the pirate gang stayed onboard the *Tungchow* while the others cast off on the junk and travelled to the mainland to unload the loot. When returning to pick up the remaining desperadoes and what was left of their ill-gotten haul of swag, a dull hum could be heard in the skies above. The unmistakable sound of an aircraft approaching sent the pirates into a frenzy as the *Fairey Seal* reconnaissance plane, launched from HMS *Hermes*, came into view. The sight of the aircraft was too much for the chief pirate, who gave the order to cut the ropes holding the junk to the *Tungchow* and ran for it, leaving his six comrades still onboard. The remaining pirates were panic stricken; realizing they were in a bad situation they started to grab hostages. The first mate as well the wireless operator along with some Chinese passengers were held at gunpoint as one of the ship's lifeboats was lowered. All the hostages were forced aboard it and, with the pirates, cast off to row hard for the mainland. Shortly after their departure a second plane, again from the *Hermes*, flew low over the ship to receive three heartfelt cheers of liberty from all those left onboard.

Once ashore, the hostages were quickly released without harm and were picked up later by a second of the *Tungchow*'s lifeboats, the pirates rendering the first useless by shooting a hole in it before scuttling off to make good their escape.

With everyone safely back onboard, the ship weighed anchor, fired up her engines, and set sail out of Bias Bay and headed for Hong Kong. It wasn't long before the Royal Navy destroyer HMS *Dainty* came in to view over the horizon and with her came the realization that the terrifying—yet exciting—episode was finally over. To quell any fears of further attacks and to ensure the entire pirate gang had left the ship, an armed guard was dispatched and placed aboard for the remainder of the journey.

The *Dainty* escorted the *Tungchow* into Hong Kong harbour to receive both the cheers of the crowds gathered at the quayside, and many sighs of relief from anxious, but jubilant, parents as they waited for the ship to dock and get the first sight of their children.

The incident, although not uncommon in the South China Sea, had captured the world's attention due to the concern for the children's safety. For a few days the press speculated what could have happened to the captives, as the notorious Bias Bay pirates were not known for their clemency. The successful rescue was a huge relief not only for the parents, but also the authorities whose job it was to ensure such incidents didn't occur. Indeed, questions were asked in the British Parliament regarding the safety of travel and trade in and around the waters of Hong

Kong, and subsequently a private parliamentary enquiry was set up to investigate the problem of pirates in the South China Sea.

After the fearful days aboard ship, the Dinsdale children were again reunited with their parents. The frightening experience was also a great adventure and one that would stay with them their entire lives. Tim, even then, had an eye for the dramatic and took his first steps into journalism by writing an account of his ordeal and entering it into a competition run by one of the local newspapers. He received second prize for his efforts:

Bleeding Like Blazes

I was walking along the deck at about ten to six, on Tuesday, when suddenly I heard a shout and somebody dashed past me with a revolver. Somehow I knew it was a piracy at once so I ran into the engine room where the second engineer was, then in to the smoking room where everybody were putting up their hands.

After that we had to stay in there for an hour and a half, with the pirates going in and out of the room. First they pushed Mr. Roxburgh and began searching him and swearing, to see if he had any pistols, the next to come in was the guard and he was searched and punched.

Once a pirate came and opened the door showing us a box of cartridges and tapping his rifle pointed at us, and the pirate next to him had a bullet through his hand and it was bleeding like blazes.

At about the same time there were about 8 or 9 shots, and plenty of scuffling and shouts. I never knew I had such a hatred for guns before in my life. Most of us were frightened like anything.

After a time we were led out and searched, then led us into the saloon where I saw the Russian guard who had put up a fight and got shot, lying there outside the door. When they went in they were searching and questioning the chief officer. They told us to sit down and eat but we didn't feel much like it, so they told us to stay downstairs.

We stayed downstairs for the night and in the early morning I and another boy named Wilkie popped our heads out to see if there was any pirates sleeping outside, and we found there was about one or two. That day passed terribly slowly, and only once a pirate came in with his pistol, and at once there was almost a dead silence. Mr.

Duncan was very kind in reading to us out of *Sherlock Holmes*, which was very interesting.

All the same the time dragged rather badly.

The last of all the days was the worst of the lot, because they were leaving the ship. I wondered if I would ever see another day again. Well we got into a dreary looking bay where the pirates were going to leave, when they said they must capture a junk to take them ashore in, and also not to be frightened if we heard any shots. When the pirates tried to capture a junk, it was outside of a lighthouse and the lighthouse signalled to some destroyers to come.

At last we got a junk and while they were getting in an aeroplane came over, and the captain pirate and three others cut the rope of the junk and sailed away leaving the others on board. We thought we might get shot, but they got off all right but took some of the officers, and left them on the beach. And we got them back all the same.

—T.K. Dinsdale, 10 yrs old.

Chapter 3

Captive

In 1936, Tim, now aged twelve, returned along with the rest of the family to the UK on a year's leave. Once in England, the Dinsdale boys became boarders at the Kings School Worcester and Felicity attended the Alice Otley School, also in Worcester. Dorys and Felix enjoyed the break, but when the time came to return to China the following year, Dorys opted to stay on in the UK a while longer and help the children settle into their new surroundings and school routine. The plan was for Doris to follow Felix out at a later date.

Securing return passage to China however proved to be not so easy. The civil unrest continued and the ongoing guerrilla war with the Communist Party was unrelenting. Furthermore, the Japanese were arming and eyeing China's natural resources and vast supplies of materials as a worthy prize for invasion. Pockets of fighting were being reported and expats were actually being encouraged to leave the areas worst affected. In spite of these difficulties and dangers, Dorys's persistence paid off with her finally departing on November 26, 1938, eighteen months after Felix's departure.

The journey, and what she found on arrival, is eloquently described in a letter written by Dorys to her sister Win, which was later printed in the local newspaper:

A Good Spot—From Which To Stay Away

After two and a half years' absence and three abortive attempts to return—each time receiving cables telling me to cancel my passage, as women were being evacuated instead of returning to the spot

28

where my husband was stationed—I at last started on my way back to China. The ship was quite full, but I was unaware of any friends or acquaintances being on board, and anticipated a rather lonely trip out. As it so often the case, I soon found various old friends, and friends of friends.

The voyage for the most part was very rough, wet and cold to begin with. The passengers seemed to consist of the usual mixed lot, which can roughly be divided into three classes: those on the first trip out, wildly enthusiastic about the sports and general activities and tremendously keen and interested in all coasts and islands and ports of call, some even stopping up all night to see the Suez Canal by moonlight, and getting a kick out of everything; those doing the round trip, who for the most part are elderly, rather frail folk for whom England is no place in the winter months; and lastly, the old timers returning East for the umpteenth time from leave. These last, rather bored with ship life, apart from bridge, and content to be left alone with a book and their thoughts of those left behind and those they are re-joining.

As I passed through Victoria (which is the city of Hong Kong) it seemed to be literally seething with Chinese of every conceivable description: richly clad men and women in their lovely satin gowns rubbing elbows with the very poorest of the poor in all their poverty and rags. Opulent and very prosperous-looking cars thronged the streets, and large departmental stores crammed with every luxury imaginable, both oriental and occidental, seemed to be on every side.

I was whisked up the Peak at great speed by way of a new and exceedingly twisty road, my ears clicking with the rapid change from sea level to 1500 feet. Roads everywhere and huge houses seemed to have sprung up in an almost magical way since I was last there; and although we reached our destination so quickly I recalled the old method of the sedan chair or rickshaw with a faint regret. In fact, it was only as I listened at night from my bedroom window to the sing-song of the servants' talk and the click-clock of the amah's[1] clogs that I could quite believe I was back once more in the same Hong Kong I knew so many years ago.

1   A housemaid or nanny.

The Refugees

I spent some days at Fanling—the golf club over on the mainland—where I had my first sight of Chinese refugees—and a pitiful sight it was. Families coming in from the country carrying their all in baskets on the ends of poles. The women carried these almost bowed to the ground with the weight of all their worldly goods piled on top of each other; babies, chickens, bedding mixed up indiscriminately. The men I saw seemed content to carry one small bundle only. As the Canton-Kowloon railway does not run past the British frontier there is a lot of disused rolling stock, and been given over for the housing of refugees—hundreds of them.

I left Hong Kong at 8 a.m. on a small river gunboat for my future home in Canton, and arrived at 4 p.m. The trip up the Pearl River was very lovely and exceedingly interesting to me as it was my first experience of travel by gunboat. As soon as we were clear of Hong Kong harbour the crew had a gun-station drill, and we dodged about, and guns were swung around; then the men got into rubber clothing and gas masks; all very interesting to the ordinary civilian. The Pearl River has been closed to ordinary merchant shipping for nearly four months, since October 1938, and the comings and goings of gunboats very restricted.

When we got to [the] Bocca Tigris forts[2] we could see the hillsides all scarred by the bombing which had taken place early on; and as we neared the boom across the river a Japanese motor boat appeared and led us through the passage. The river is very wide and shallow at this point, and in spite of various craft—steamers, junks and sampans sunk by the Chinese—there was really plenty of room to pass through.

Approaching Canton, gaunt, skeleton buildings completely gutted by fire stood out against the skyline, giving a desolate aspect to the countryside. When I got to our flat there were many evidences of the terrible explosions which shook Shameen[3] on October 23. In our

---

2 Bocca Tigris is a narrow strait in the Pearl River Delta littered with islands, some of which were fortified. The region hosted a number of battles during the First Opium War of 1839–42.

3 Shameen Island, today known as Shamian Island, was where European traders settled and did business thoughout the eighteenth and nineteenth centuries. It was of strategic importance during the Second Opium War (1856–60), after which it was divided between the two colonial powers of Great Britain and France.

dining room alone paper pasted over gaps replace 15 panes of glass.

I was taken to the city on one of my first days here, and a more desolate and deserted place cannot be imagined; miles and miles of streets completely devastated by fire, ruin everywhere, electric cables hanging, and poles leaning crazily; window frames swinging loose by one hinge and filth and rubbish everywhere, and hardly a soul about in the usually thronged thoroughfares. A terrible example of modern war and a most sad and depressing sight. On Shameen, food, coal and wood are controlled, and we even have 12 cows on the island.

Foreigners and missionaries of all denominations organized refugee camps—20,000 were being cared for at one time—taking over hospitals and lunatic asylums, from which the staffs had not only fled, leaving their patients with no one to care for their simplest of needs, but had taken everything they could lay hands on in the way of medical supplies as well.

Conditions in the city and surrounding districts are very bad, looters and armed robbers terrorizing the whole countryside. Commerce is at a standstill, and no one predicts when it will be better.

Social life on Shameen goes on as usual with its round of sports—tennis, bowls, badminton, football and hockey—the latter two games played chiefly by the men off the gunboats. In spite of the war there seem to be as many dinner and cocktail parties as ever, and apart from aeroplanes overhead, occasional gunfire in the distance, and bumps of explosions, we hear little, and know less, of what is happening. Lots of rumours but no authentic news.

My household goods, stored in Hong Kong, come a case or two at a time, if and when a gunboat comes and can bring them. Letters are very few and far between—coming and going also by gunboat. It is two months since I left England, and already China strikes me as being a very good spot—from which to stay away!

With the threat of all-out war between China and Japan becoming a real prospect, Dorys departed for England in May 1940 travelling on the Blue Star line heading for Southampton via South Africa, whereas Felix moved to the comparative safety of Hong Kong. Britain and Japan at that time were still on friendly terms, and with Hong Kong being a British colonial territory the colony was viewed as a relatively safe place to stay and continue business.

However, that all changed on the morning of December 8, 1941, when, just a few hours after the unprovoked attack on Pearl Harbour, bombs started raining down on Hong Kong's Kai Tak airport. The Japanese had turned their focus on gaining access to the oil fields in the European-controlled area of Southeast Asia. They declared hostilities with France, Holland, and Great Britain.

The battle for Hong Kong lasted seventeen bloody days. A combination of British, Canadian, and Indian troops led a spirited yet futile defence against a far larger Japanese force. Outnumbered three to one and with no airpower, the coalition, realising the situation was hopeless, was left with little choice but to lay down arms and save further loss of life. On Christmas Day, 1941, Hong Kong became the first British Crown Colony to surrender to an invading force.

Much has been written about those first days of lawlessness of the Japanese takeover. It is well documented that the occupying soldiers became drunk with their success and rampaged through the streets where a complete breakdown of law and order had occurred. During this period the non-Chinese civilian population, numbering approximately 3,000, were, for the most part, left alone and told to stay in their homes for their own safety. Shortly afterwards, however, all civilians were rounded up and placed in an area known as the Stanley District, which was to become their prison for the next three and a half years.

Stanley Internment Camp, as the site became known, was where Felix spent the rest of the war. Along with all the other internees, his daily camp routine was, first and foremost, to secure enough food to survive until the next day; the Japanese provided a pitiful amount of supplies that, more often than not, were already rotten or contaminated on arrival. Malnutrition, disease and sickness became rife, and starvation ever present. In order to supplement the meagre, almost inedible, rations, a thriving black market developed in which food and other vital goods were brought into the camp for sale or trade by the guards. Prices were exorbitant and to pay for the much-needed—indeed life-giving—supplies, everything was sold. In a letter to Dorys, Felix explains the sale of both his wedding ring and watch to raise funds to buy much-needed food:

"Needless to say, it was with extreme reluctance and under the urge of hunger that I parted with the [wedding] ring and watch but I feel in the circumstances you will think me justified, and I hope this is so. The money they brought in has been a great help."

The black market economy in Stanley Camp was credited with the relatively low death toll, considering the conditions. Survival of many of the internees was directly due to the enterprising efforts of the "wheelers and dealers" who kept a steady flow of goods coming to the camp.

The Japanese surrendered in August 1945 and Hong Kong was passed back into the hands of the British. Felix had survived the years of internment but at the time of liberation was extremely ill. At five feet ten inches he weighed less than ninety pounds and was racked with tuberculosis. The doctors didn't expect him to survive the long journey back to Britain, so he was instead placed on a ship departing for Sydney, Australia, for transport to the specialist TB hospital there where he'd get excellent medical attention quickly.

After a short stay in hospital, Felix rallied. His health improved just enough for the doctors to give the okay for him to travel, and later that year he departed on a ship bound for England. A cable was sent to let Dorys know that Felix was indeed on his way home, and with it came excitement and a hope that, after all the years of worry and fear brought by the war, finally the family would once again be reunited.

However, that dream wasn't to be realized. Less than twenty-four hours into the voyage, Felix suffered a severe haemorrhage and died on board the ship. He was buried at sea just off the Australian coast.

The news was devastating. After the years of anxiety about the dreadful conditions at the Stanley Camp he'd had to endure, his liberation and his rapidly improving health had made the family elated. To learn he'd succumbed to his illness when he was finally on his way home was a cruel twist indeed.

Dorys never fully recovered from the shock of Felix's loss; she lived to the grand age of ninety-two but never came to terms with the seemingly unjust way things had happened.

Chapter 4

Canada's Haunting Chill

For the Dinsdale boys, private school wasn't an easy ride; the education was good, but the discipline strict. In those days, the compulsory cold showers after any sporting activity—no matter what the weather, the season, or your state of health—was a torturous act, said to "toughen the boys up." The pupils hated it. Indeed, later in life Tim always maintained his "dicky ticker" (weak heart) was primarily due to this awful practice.

With both their parents thousands of miles away, it took eight weeks for letters to reach the children. It was a lonely schooling, with the three Dinsdale children only getting to see each other at church on Sundays. There were, however, always the school holidays, which were not long enough to be spent with their parents in China but certainly enough time to head off to Wales and their wonderful Aunty Win's cottage on the beach at Borth. The summer months were filled with adventure and the children spent hours playing in rock pools, catching prawns, searching for new fishing spots, building huge sand castles and waiting to see whose would fall last when the tide came in. Borth became their home away from home and Aunty Win became their loving and caring surrogate mother. Those long summers away from the strict, regimented school life were cherished and eagerly anticipated each year; but as each summer passed the atmosphere in the country was changing. With the onset of war, the country's youth was, once more, asked to stand up and protect the liberty of the British nation.

When the Japanese interned his father, Tim wanted to get fighting in the Far East as soon as he could. As with many public school boys of the time, the allure of the air was strong. His desire to fly was certainly helped by an incident on the school rugby field when a lone Spitfire flew low over the school, barrel-rolling and doing aerobatics, swooping down to "buzz" the boys on the playing field, and generally showing off to all the cheering fellows below. A spectacular sight and of course boy's own stuff. It was later, when they found out the pilot was a past head boy from the school, that Tim became really hooked; he set his sights on a future flying fighter planes with the RAF, ideally, in the war raging in Asia.

School was coming to an end and the RAF was where he eyed his future; however the first port of call after leaving Kings School at age eighteen was the de Havilland Aircraft Company based in Hatfield, Hertfordshire. There Tim not only started his career in aviation as an aeronautical student in 1942, but also his military service, joining the local brigade of the Home Guard. The country was at war and the threat of invasion very real, so able-bodied males of all ages joined the Home Guard. The intention was to defend their homes, villages, towns, and cities should an attack come. Of course, as with any military unit, soldiers need training, and it was on a training session, conducted with live ammunition, when Tim sustained his one and only wartime injury by taking a bullet to the hand. It was a pure accident but the bullet lodged between his fourth and fifth finger on his right hand where it was to stay for the next twenty-eight years.

Gaining entry to the RAF proved to be a little harder than Tim had originally hoped. On his first attempt it seemed testing in low-level light and at distance had been a problem—a bit like taking your driving test on a dull day and having to read a car number plate at sixty yards. After the disappointment of the first failed attempt, Tim reapplied and got a second chance. His experience the first time round was to prove invaluable when an opportunity to "help himself" was presented. One test had all the budding pilots sitting in a line in a darkened room. Their goal was to identify targets as they came on display. To stop anyone getting a closer look, a small length of string was tied to the back of everyone's chair; the string had a hook attached which was placed on the back of each student's collar thus stopping the students from leaning forward. Tim worked out that the examiner walked up and down the line checking that each hook was in place just before the new target came up. Once his hook had been checked he'd quickly slip it off, lean forward, identify the target, and replace the hook before the examiner returned. Whether it was this little bit of self-determination which made the difference or not, it didn't matter as Tim passed the entrance exam and was accepted to the RAF flying school. His wish to become a pilot and fight for his country was becoming a reality.

Britain needed pilots badly, so flight school commenced immediately. It was a tough schedule split between the classroom and getting off the ground to learn the basics of flying. The pressure of flight school was especially intense because many of the young men would be flying in combat in the not too distant future, and the stark reality of that was "kill or be killed." When the opportunity arose to continue his flight training in either Canada or South Africa (the RAF had a programme in which cadet pilots were shipped off to different parts of the Commonwealth to continue their flight training) Tim jumped at it, opting for South Africa as his preferred destination; Canada, he was sure, was a country he'd most likely visit after the war. Heading south this time would mean getting away from the dreary cold of the northern hemisphere winter.

The troop ship sailed from the River Clyde in Glasgow destined for South Africa in the winter of 1944. It wasn't a particularly pleasant journey and certainly nothing like the Far Eastern voyages of his childhood.

In Tim's own words, "it was cold and miserable as we shuffled aboard, loaded down with kit. Deep into the bowels of the ship the column inched forward like a caterpillar, and at a place where there was scarcely room enough to stand the order had come to 'make yourselves at home.' It was our mess deck, the space in which we were to exist for a month, and through the stifling heat of the tropics. It was trooping at its wartime worst, and the introduction to it—the [river] Clyde, dank and dour as I saw it then—was something I wanted to forget."

Far from the chill of the northern winter and the shortages of wartime Britain, Tim's flight training continued. The stunning beauty of both the South African and Rhodesian[1] countryside and the near perfect weather made flying conditions exceptional. The experience, however, was to be short-lived; the war was coming to a rapid conclusion in Europe, and although Tim's desire was to go on and fight in the

---

1 Rhodesia, a British colony named in 1895 after the founder of the British South African Company Cecil Rhodes, gained independence in April 1980 and was renamed Republic of Zimbabwe.

Far East, the powers that be decided otherwise. The pilot programme was shut down in the Commonwealth countries, with all the cadets returning to Britain to await their turn to be demobbed.[2]

Although Tim only had a short time in South Africa he made some good friendships. One friend in particular was to remain close to the Dinsdale family for the rest of her life by becoming my sister Dawn's godmother. Gladys Marshall, a lovely lady, had helped the war effort by billeting some of the cadet pilots during their training. She housed and fed the boys and generally took care of them, becoming their mother while they were away from home. It was a kindness Tim never forgot.

Demobilized from the forces with his pilot's licence in hand, Tim's only train of thought was to complete his aeronautical engineer's apprenticeship at de Havilland. He knew his future was in aeronautics. The industry was breaking new ground with jet aircraft set to take over from the wartime propeller-driven fighters and bombers. The industry was exciting and innovative and Tim wanted to be a part of it. He returned to de Havilland, where at the time the young apprentice technicians and engineers were housed at Digswell House[3] in Welwyn, Hertfordshire, a classic Regency-style country house the company owned.

In those early post-war years, Digswell House was the scene of much merriment. The war was over and they had survived. The young men took every opportunity to make the most of their freedom: weekend dances were commonplace, as were trips to the town's hot spots to swill back a pint or two and hopefully catch the eye of an attractive girl.

Shortly after graduating from technical school, and now working full time in the aircraft industry, Tim was invited to the local Blue Harts Hockey Club annual St. Patrick's Day dance in Hitchin, Hertfordshire. There he was introduced to the young and charismatic Wendy Christine Osborne. She had the biggest smile in the place and the energy to dance anyone off the floor. Tim was smitten.

It was a whirlwind romance. Tim proposed seven months later and Wendy became Mrs. Dinsdale on June 25, 1951 at St. Mary's church

2 Demobilization, a process where soldiers are discharged from the military via a system of service and age.

3 In 1957, after de Havilland left, the Digswell Art Trust was organized in order to restore and preserve this fine example of Regency architecture. The house, grounds and out buildings became art studios and artists' accommodations, a purpose to which it continues to serve more than fifty years later.

in Hitchin town centre. After a honeymoon in Wales, Tim was quickly off to Canada to take up a position with Avro Aircraft Ltd. based in Toronto. As he had foreseen years earlier, Canada was becoming the world hub of the aircraft industry, and talented young designers, engineers, and pilots who wanted to be at the cutting edge of aviation development were heading there. Wendy followed a month later. Having barely left Hertfordshire before this, she was excited but apprehensive about making the trip alone; after all, Canada was half a world away. Her parents were supportive, but one can imagine their worry as their nineteen-year-old only child stepped on the train to depart for Liverpool and ultimately Canada for who knew how long a time.

The start of the six-day crossing was delayed when the former wartime troop carrier, RMS *Franconia*, ran aground in the River Mersey while departing from Liverpool. Once the tide turned, the ship floated easily off the mud bank and started on her journey. However, the voyage was rough, the North Atlantic putting on a show for the novice sailor with twenty-foot swells, conditions that kept Wendy firmly locked up in her cabin, wracked with seasickness. After a few days, a friendly crewmember suggested a trip up on deck to get some fresh air would help, so Wendy made her way up to attend the Sunday church service. After sitting for a while and starting to enjoy the fresh chill of the sea breeze, the congregation started to sing the hymn "For Those in Peril on the Sea." This was enough to send an already fragile Wendy back down to her cabin, not to be seen again until the ship left the turbulence of the open sea and entered the calmer waters of the St. Lawrence River.

Tim went to meet his young bride and whisk her back to Toronto where they were to start their new adventure. Life in Canada proved to be quite different to that in post-war Britain. First off, living in downtown Toronto in August was a shock for both Wendy and Tim. The humidity count was off the clock. It was a heat the likes of which Wendy had never experienced before, and although Tim's Far Eastern upbringing had exposed him to the same kind of intensity as that of an Ontario summer, it was many years since he'd left, so he too suffered from the stifling temperatures.

The winters were also a complete contrast to those in Britain, where, for the most part, it just rains a lot more than it does in the summer. Their first Canadian winter was an eye-opening experience. Months of sub-zero temperatures coupled with several feet of snow meant doing everyday things had to be relearned. They had to learn to drive in snowy conditions and make sure there were enough supplies in the house, as "just popping around the corner to the shops" wasn't an option. Having the right clothing was imperative, as an English winter coat was no match for the arctic chill of the Canadian winter months. But they adapted and started to enjoy their new surroundings. Tim took up skiing, trying some cross-country adventures. He also tried his hand at ice-skating, but quickly decided a future playing in the NHL probably wasn't on the cards. Wendy got a job in the flight test centre at Avro and traveled to work with Tim in his aptly named Oldsmobile each day. Before long they started to develop a group of friends and settle into the Canadian lifestyle.

One of the more charming Canadian traditions was to build a summer cottage on a lakefront property. With a bit of searching and some good luck, Tim and Wendy joined a small group who purchased a piece of land in a glorious setting on Lake Kawagama in Haliburton County, about two hours' drive north of Toronto. They built the cabin from scratch, ferrying all the materials over by boat, as the area wasn't accessible by road. It was a beautiful secluded spot where they would spend many summer weekends enjoying the sunshine, swimming, and water skiing. However, Wendy, a good Hertfordshire girl, never really got over her fear of bears, and, as they were in slap bang in the middle of bear country, insisted on being accompanied on her every visit to the outhouse.

In 1952, Tim advanced his career with a move to Rolls Royce's aircraft division in Montreal. Wendy, now pregnant with their first child, Simon, relocated to a large old house in the Beaurepaire area. It had a wraparound veranda and a view across the frozen St. Lawrence River, but for all of the house's period charm, Wendy didn't like the place one bit. Being left alone all day was one thing, but, moreover, she felt uncomfortable there and became convinced there was something strange about the place. Ghost stories cut no ice with Tim; in fact, he found the whole thing rather annoying as he felt perfectly calm and at ease there. It wasn't until a couple came to stay for a few days that Tim gave any credence to Wendy's concerns. After dinner one night, John, their guest, leapt out of the chair and went over to the bottom of the stairwell. He was looking up at the curtain which he'd just seen swish from side to side as if being brushed by somebody walking past. When quizzed what the problem was, John, ashen faced, dismissed it as being the wind moving the curtain. However, as Wendy pointed out, the storm shutters were closed and it was a perfectly still night!

A few weeks later, a second incident occurred. Another friend had stayed overnight, and when he left, unaware of any of the previous stories, he mentioned that he had a very uneasy feeling in the house and felt there was something "strange" on the stair landing—the same spot where John had seen the curtain moving. Wendy decided enough was enough, and it wasn't long before they vacated the house and found newer accommodation with a more central location and a cosier atmosphere.

Another career opportunity came Tim's way in 1956 and with it the prospect of moving back to the UK; it was a good job offer and a move in the right direction on the corporate ladder. Plus, the family was expanding, Alexandra had arrived two years after Simon, and so the chance of cementing his career couldn't be missed. Additionally, the children hadn't seen their grandparents and Wendy was missing her folks too, so the decision to pack up and leave Canada and return to the UK was an easy one to make.

Chapter 5
The Monster's Grip

The Britain they had left five years earlier was much changed. The country was mending after the destruction and shortages of the war years. Industry was thriving, there was full employment and, along with a new Queen, the hopes for a brighter future were high. The one blot on the landscape was the disastrous Suez Canal crisis, but that was relatively short-lived, and with Prime Minister Anthony Eden resigning over his gross miscalculation, the country soon regained its footing and continued on a path of rebuilding.

Tim purchased a modern semi-detached house in a new suburb just on the outskirts of Reading. It was a pleasant, three-bedroom home with a nice-sized garden. Wendy, enjoying her first real home, wanted to plant a magnolia tree in the front garden, but Tim wasn't keen, as magnolias are slow growers and with the family expanding once more they probably wouldn't be staying there long enough to enjoy the benefit. A compromise was made; an English rambling rose was planted instead. Over fifty years later, and still in the same house, the rose continues to festoon the front porch, blossoming each summer and giving pleasure to Wendy and her neighbours.

It is at this point when our story takes an amazingly sharp turn off the beaten track. So far Tim had led a pretty average life; an interesting life, yes, with all his Far Eastern experiences—and certainly being pirated was anything but normal—however a solid professional career in aeronautics and a young family of three really wasn't anything out of the ordinary. This all changed one day in March 1959, when Tim was sitting at home reading a popular magazine of the time, and a particular favourite of his, *Everybody's*, and he came across an article titled, "The

Day I Saw the Loch Ness Monster." As today, it was a subject most folks seemed to have heard of but knew very little about. Indeed the fabled monster was more often than not attributed to the Scottish mist or the result of too much Highland "spring water" (whisky) than given any credibility.

Tim read the piece with a mild interest. There was a blurry black and white picture showing "something" on the loch's surface and a number of eyewitness accounts which varied in their descriptions but, nonetheless, all talked about a very unusual and, as yet, unknown creature that seemed to inhabit the loch. The article gave some background as to the history of the legend, stating that strange, unexplainable sightings of large animals had been going on for centuries and the myth of the *eigh-uisge*, Gaelic for water horse, had a strong association with Loch Ness.

Passing the magazine over to Wendy, Tim asked her to read the article and waited for her opinion. Wendy, practical as ever, said, "I think there's probably something in it. These tales of a monster have been going on for years and years, so there must be something unusual in the loch, or there wouldn't be so much fuss about it." Tim was in agreement. Why would seemingly ordinary folk want to lie about such a thing?

Yes, there had been a well-documented hoax in the early thirties when some large, unusual footprints where found at the side of the loch. An experienced big-game hunter examined the prints and stated he'd never seen anything like them before, and so interest grew. Plaster casts were taken and, amid much excitement, shipped off to the British Museum where, on analysis, they were found to be nothing more than the prints of a female hippopotamus foot—probably the dried remains of a hunter's trophy. Such episodes created an aura of eccentricity and the ridiculous around the subject; as a result, the topic of the monster was nearly always spoken of tongue-in-cheek. This mind-set meant folks who would openly report a sighting could expect some heavy ridicule. This all brought Tim back to the same question: Why lie? Why on earth would people make up such amazing stories knowing full well they would most likely get mocked, laughed at both publicly and privately, and for the most part not believed? It was a puzzling question, and to Tim the answer could only be that they must simply be telling the truth.

Reading the article a second time and mulling over the individual reports it featured, which spanned decades, Tim felt an interest growing; in fact, it was more than an interest, it was a curiosity, the engineer's brain wanting to know more. He took out a pen and paper and started to analyse the accounts, jotting down points of interest, similarities between sightings, the animal's behaviour, colour, the time of day it was seen. In fact, anything that could be collocated and studied, he did. After this rudimentary exercise he categorized his conclusions into five areas of consideration:

First: of the ten or so sightings recorded, the majority seemed to have occurred in the very early morning, just after dawn, suggesting the animal might conform to a routine or be nocturnal in habit; this could also explain why it was so rarely seen, although on the basis of a few reports it would be a mistake to draw any firm conclusions.

Second: people—completely disassociated and reliable people—seemed to say much the same about it and indicated surprise and astonishment in terms that were both simple and candid; and this suggested truth.

Third: notwithstanding the promise of Sir Edward Mountain's expedition in 1934, no official expedition had ever visited the loch in all the years that followed, in spite of the continued flow of evidence.

Fourth: hoaxes, confusing though they may have been, had generally been exposed as such, and did not appear to explain the hard core of evidence.

Fifth: due to the relatively narrow width of the loch, averaging a mile and a quarter right down its twenty-four-mile length, the surface could be put within range and scan of a number of telephoto cameras, mounted at points of vantage on either side.

Little did Tim know at the time, but that hour he'd invested in doing the simple analysis was to be the spark that would ignite his imagination and set him, and his family, on a life-changing path of mystery, intrigue, and adventure.

His conclusions did nothing to quell his curiosity. On the contrary, the deeper he delved into the subject the more intriguing it became, and more and more unanswered questions came to the surface. Questions such as: why, after the long history of something very unusual living in the loch hadn't there been any proper research undertaken by an official body, a university or a government-backed project or perhaps the British Museum? And why, with the animal's consistently reported great size, wasn't it seen more often? And again, why hadn't any remains of the creature been found? These and many more perplexing questions swirled around Tim's mind, gripping his thoughts and, over the weeks that followed, refusing to let go.

He thought about Sir Edward Mountain's self-funded expedition of 1934 and how his game plan had been to position a number of paid watchers at various vantage points around the loch. After a month of daily surveying the surface waters from morning to night the results were less than conclusive. Only a few feet of ciné film had been shot, producing nothing that could be substantiated. This failure didn't seem to concern Tim, as he felt sure the concept of many eyes watching the loch at the same time was a good one. So, fuelled with enthusiasm, he set about devising a plan that would lead to just that.

A few weeks later and Tim's first Loch Ness expedition strategy was ready. He produced a letter outlining the project, with timelines, costs, and expected results, then set about gaining interest and financial support from a variety of areas. Government departments, industry, and philanthropists were all contacted. Tim waited, and waited some more. The response was deafening in its silence. Whether the plan was too ambitious, or the subject matter too quirky, it seemed nobody wanted to risk either reputation or money on such a scheme.

Disappointed at the lack of interest, but unperturbed, Tim set about modifying his plan and at the same time realised he needed to increase his own meagre knowledge of the subject. The book, *More Than a Legend*, by Constance Whyte gave him greater insight into the myth surrounding the monster. Constance had done an excellent job at compiling many years' worth of sightings; as the wife of the Caledonian Ship Canal[1] manager she had heard plenty of monster stories over the twenty-three years her husband ran the canal, and had talked to a number of people who claimed to have seen the creature. This new source of information just spurred Tim on. With the realisation that the subject was far larger than he'd first given it credit, he decided to adjust his approach and take a more scientific view of the whole topic.

Now, with the hooks of the mystery firmly embedded, Tim set himself a goal of researching one hundred (at a minimum) monster

1   Opened in 1847, the canal connects the Scottish east and west coast via a sixty-mile combination of lochs (lakes), locks and canals.

sightings. His mind—that of a trained engineer and so by consequence logical—was open to whatever he might find, whether it was just a number of elaborate hoaxes, people playing practical jokes, or perhaps there really was an extraordinary creature, new to zoology, living in the loch. With an impartial approach, Tim set out to collate and examine all the reports he could lay his hands on. He developed a chart where he categorized the information into tables, thus building a picture of specific points drawing similarities between sightings and reports.

It was a giant undertaking. Tim was doing his homework and in doing so becoming well informed in the field of Nessie. What had started off as a passing interest when reading a magazine article had become a serious and compelling study of an unknown colony of animals living in Loch Ness.

During all of this, Wendy showed mild interest, but with three small children to care for she had little time for fabled monster stories and, for the most part, left Tim to get on with his studies once the children had all gone to bed. This quiet time was spent poring over newspaper clippings and reports of sightings that people had sent him. It was a five-month undertaking, an incredible journey of understanding, amazement, and, at times, disbelief as an incredible image of something unique started to appear before his eyes.

Upon analysing his own study, Tim encountered the shocking nature of what the evidence was telling him: an animal of up to forty feet in length, with large diamond-shaped paddles, a long neck with a small, almost reptilian-like head, a powerful tail that could churn up the water like a paddle wheeler, and, most curious of all, the humps—presumably the animal's back—which was the one thing about which the reports varied tremendously. Some eyewitnesses reported one, two, or even three humps, while others only mentioned seeing one long back—like an up-turned boat. But a point all agreed on was the tremendous speed the animal could generate. This indicated a creature of great strength.

It was stunning stuff. A hundred reports studied and the similarities were startling in their consistency, other than the number of humps, which in Tim's eyes only added to the authenticity of each account. He rationalized that if people were making it up why would they differ on such an important point? It didn't make much sense other than people were telling the truth. This then, of course, brought up yet another intriguing conjecture: those varying numbers of humps, along with the other astonishing conclusions regarding the physical make-up and shape of the creature made it appear that it might even be able to change its body shape, an anomaly not unheard of in the animal kingdom. But it was speculation, and that was the one thing Tim was determined to stay away from. By sticking to the facts, as he'd deciphered, he wouldn't get lost on a path of speculation. However, saying that, it was an intriguing possibility nonetheless.

Tim wanted to give the monster "life." Using his study as the blueprint, he set about constructing a clay model. It wasn't long into the project before he realised the form starting to take shape was nothing like that which existed, or was known to exist, in the world. The descriptive words of the eyewitnesses were now taking a tangible shape, and with it a distinct appearance, something almost prehistoric, was materialising.

After months of building a picture, and now a model, the next move was an obvious one; Tim needed to visit the loch. He had seen, read and analysed enough evidence to be convinced there was something very real and extraordinary inhabiting Loch Ness. He'd done his studying, so now wanted the field trip. A visit to the loch would mean getting a feel for the surroundings he'd read so much about. There would also be the possibility of meeting witnesses—folks who had come face-to-face with the legendary monster—and getting the human emotion and feeling only a firsthand account can provide.

It was an exciting prospect; however, the expedition would have to wait a while longer. Now with his research complete it was winter and not an ideal time of year to be travelling so far north. A plan was put in place to visit the loch in April and spend five days watching from various vantage points and meeting locals with knowledge of the animal—real knowledge, as just about all the loch-side residents had a monster or "beastie" story of one kind or another to tell.

Introduction letters were written, equipment accumulated and the hotel booked. The car was modified to accommodate all the expedition needs, and a deep breath drawn. Tim was leaving seven-months-pregnant Wendy alone with three small children to go off and chase a mythical monster in the far north of Scotland. He questioned his own reasoning, but with Wendy's reassurance and blessing, turned the key and fired up the engine, and with its ignition came a flame of excitement about what lay ahead. It was just over a year since he'd first read the article of strange sightings in Loch Ness, a subject about which he had no previous knowledge, and now twelve months later he was practically an expert.

As Tim pulled out of the driveway on that fateful April morning, waving goodbye to Wendy and the children, he couldn't even begin to guess what lay ahead. He knew he was on a great adventure, but the magnitude of how life changing it would turn out to be wasn't even a thought.

Chapter 6

A Cameo Appearance

At this point I'll turn over the story to be told in Tim's own words. The following is taken from Tim's first book, Loch Ness Monster. It is his dramatic account of his first expedition and subsequent events.

On 16 April 1960 I set out from my home in Reading in a small car loaded to the roof with equipment. I knew that the months of work and planning lay behind, and that if I did not find the Monster I could at least fulfil a number of other useful tasks at the loch, which would add to my general fund of knowledge and perhaps prepare the way for future expeditions.

I was on my own, because in spite of making every effort to persuade others to come along, both friends and relatives alike had declined the invitation! Business matters, and the chilly prospect of days of fruitless searching, spent out of doors in the rain, perhaps, had brought a rush of polite refusals; but I was not perturbed, and the prospect of attempting to search the 14,000 acres of water unaided was very much a challenge.

During the planned five-day stay at the loch I had to accomplish three things; look for the Monster, each day, from dawn till dusk; carry out a physical reconnaissance, and map those parts of the loch which were best camera sites and camping places; and in the evenings, talk to local people who claimed to have seen the creatures. This would make a very long day and if I hoped to be successful, I would have to maintain mobility and a state of instant readiness, regardless of wind, or rain, or bodily fatigue. It would require a determined effort, and yet it was not without promise of reward. From end to end the loch is relatively narrow and with the telephoto lens I had on loan, I could reach out and film the Monster up to a range of a mile or more—if only I could see it.

Driving up through the industrial heart of England I joined the Great North Road, and late that evening drove through the gates of a Northumberland farm, glad to have completed the first leg of the wearisome journey. The following day I took the route up through Edinburgh, across the Firth of Forth by Queen's ferry, under the magnificent soaring spans of the old railway bridge, and soon into the mountains and glens to the north.

Looking about, as any stranger would, I began to wonder whether I had made a mistake coming so early in the year. Inside the car it was deceptively warm and pleasant, but when I opened the window a freezing wind swirled about, bringing with it a taste of the rigours to come.

There was little traffic on the road which wound its lonely way ahead, far into the distance. My spirits began to sink and for the first time in months I was plagued again with doubt. Was it all a fairy tale after all? Could there really be a Monster in the loch? What on earth was I doing driving on all alone, into this land of rock and ice? Perhaps it was all a load of rubbish, the misguided chatter of silly people! But it was too late to turn back. For the hundredth time I added up the facts and arrived at the same encouraging totals, and as I neared the journey's end I left behind the bleak and sullen mountain scenery and in place of it the Great Glen of Scotland, the huge rift dividing north and south, linking the west and eastern seaboard.

I drove into Inverness, a mellow tidy place, with shops and banks and all the outward signs of life of a modern thriving township, so different from the outpost I had imagined, and then out along the little road running parallel with the shallow River Ness, on toward the loch itself. I had several miles to go, and the light was beginning to fail, so before reaching the water I stopped the car and set up my camera equipment, determined that whatever happened the Monster would never catch me unawares.

I had three cameras with me, a tripod mounted 16mm Bolex ciné with 135mm telephoto lens loaded with black and white film, an 8mm Kodak ciné and a good German 35mm both loaded with colour film. The smaller cameras were for stand by purposes only, and would be used if I ran out of 16mm film, or had time to spare.

I had removed the left-front seat from inside the car, and a folding canoe lay in its place, and on top of that a load of other equipment. My tactics were simple enough—I would use the car as a mobile platform, and chase the Monster, should I get to see it in the distance, and thus try to close the range and obtain those vital close-up pictures. I knew the animal was supposed to be enormous, and that if it surfaced it might splash about for ten or twenty minutes. With my binoculars, I could scan the loch for two or three miles at least in either direction, with the car I could quickly close the gap, and if the Monster was then still too far away I could, in theory, launch the assembled canoe from the roof of the car and paddle off towards it! This was the temporary plan I had decided on, but I knew that it would need to be flexible, adjusted in whatever was seemed best in the light of experience gained on the shores of the loch.

Driving down the narrow road leading to the southern shore, I got a first glimpse of the loch from a point on high ground about a mile from its eastern extremity. Breasting a rise, I stopped, and there, stretching away into the distance as far as the eye could see lay a great shining pool of water, reflecting the last rays of a wintery sun, framed on either side by blackened walls of mountains, I stood for a moment gazing down at the scene, much affected by the strange beauty of the place; and then, as the light was fading, turned and climbed back into the car, conscious for the first time of being tired. In the last two days I had spent nearly twenty-five hours camped in the driver's seat, and that was more than enough.

I drove to the village of Dores, in amongst the trees, and then out along the General Wade's military road, the narrow single track running parallel to the southern shore, almost level with the water. As I drove along I instinctively craned my neck, staring out over the water, hoping vainly to see the Monster, but knowing all the while that I would not.

The road started to climb a little and the trees grew thicker—and then, several hundred yards ahead I saw a man at the roadside, peering out across the loch, pointing, and beyond him a women and two children waving their arms in excitement.

Intrigued, I drove up closer, trying to drive and look all at once—and then, incredibly, two or three hundred yards from shore, I saw two sinuous humps breaking the surface with seven or eight feet of clear water showing between each. I looked again, blinking my eyes, but there it remained as large as life, lolling on the surface!

I swung the car across the road and locked the wheels, pulling up in a shower of gravel; and flinging open the doors, lifted the Bolex out, with its long ungainly tripod. I struggled to set it up on the uneven ground, and with palsied hands set about the task of getting the camera into action. I kept glancing at the loch, expecting to see the humps disappear in a sudden swirl of foam.

I knew the seconds must be flying by, but the unfamiliar camera—a mass of knobs and levers, glinting back at me in the distance—would not be hurried. By now my hands were shaking to such an extent I could do little useful with them; but when at last, almost in despair, I squinted through the sight ready to film, the humps were still in place, calmly awaiting events: floating on the surface, strangely docile and inanimate.

For a moment I hesitated, my finger on the button, and then upon a sudden impulse reached for my binoculars in the car, and focused them upon it. Expanded seven times, the humps looked more impressive, larger than life it seemed, and yet when I examined them carefully it was just possible to see a single hair-like twig sprouting out of the one to the right—with a solitary leaf upon it, fluttering gaily in the breeze.

Slowly and deliberately I put the equipment back in the car, feeling rather foolish, and drove off up the road, watched by curious eyes of the man ahead. It was probably a good thing the Monster had turned out to be a floating tree trunk. To have filmed it too soon would have almost been a disappointment, and my first attempt to get into action quickly had not deserved success. I had made a hopeless mess of it, wasting time fiddling about in a state of wild confusion, and I realized that if I was ever to film successfully, I would need to drill myself in the use of the camera as though my life depended on it, practising every moment until reactions came with instinctive speed and complete calmness.

By now the light was fading fast, so I drove to the little hotel at Foyers perched high up on the hillside commanding a splendid view of the loch. Hugh Rowand, the English proprietor, came out to greet me, and leading the way showed me to my room upstairs, overlooking the waters, two or three hundred feet below. He seemed a very pleasant person and politely ignored the telltale cameras in the car, and I was glad that he did, a little sheepish still from my encounter with the tree trunk!

I washed and changed, and ate dinner, but before going to bed I set up the tripod and camera and worked on my drill until I was proficient. To operate a 16-mm ciné camera of the type I had with me required five separate deliberate adjustments, each of importance, and two more on the tripod. I knew that to omit any of these or to make a mistake about them might well result in a defective film—or worse, an opportunity missed forever.

When everything was set up to standard, I went to bed, having first set the alarm to go off at 4.30 the following morning, half an hour before dawn.

The hunt begins in earnest.

Daily log April 18th 1960. Easter Monday

4.30 am	Got up
5 am–9.30 am	Drove round the loch
9.30 am–6.30 pm	Survey work. Area 1 (see map)
7 pm–9 pm	Foyers Point

Results: Monster–nil. Survey–useful. Interviews–one good. Total watching time, approx 16 hours.

Behind these cryptic remarks in my daily log lay a varied and rewarding day's activities, made after a rather discouraging start. At 3 am, only an hour or two after going to sleep, I had woken again, and as the hours ticked off into the morning I lay in miserable suspense, waiting for the alarm to ring...

Outside, the loch lay cold and dim, far below and out of reach in the very early light. A few minutes before sunrise I drove off up the steeply climbing road to Fort Augustus, 14 miles to the west. Twisting and turning out of sight of the loch the tiny road climbed off into the mountains, and I followed it in a state of depression—and yet as the road climbed higher and higher, I could not help but look about in curiosity; and it was then that I became aware of the different colour tones, and a gradual change in light. I stopped the car and got out to obtain a better view, and there to the east I saw a sight of quite extraordinary beauty—a great fiery ball of sun had arisen, and all about the jagged snow-capped peaks of the mountains reflected its brilliant orange light! I stood for a moment entranced, gazing down at this wild and lovely scenery, feeling as though I had wandered into a dream. The air was fresh, and I began to realize that for all my doubts and chills there would be compensations in this lonely search for the Monster, which as yet had hardly begun. In the next five days I must inevitably spend time out of doors in country such as this, living and moving amongst the peaks, breathing this pure and fragrant air.

Much encouraged I carried on to Fort Augustus, past the little Loch Tarff, glinting at the roadside, and then down the mountain towards Loch Ness. I first caught sight of it again from a place 700 or 800 ft above the water, and realized what an ideal point of vantage it was. Below the western end of the loch could be clearly scanned, including the beach at Borlum Bay—the place where the Monster had been seen partly out of the water in 1934; but as the purpose of the journey was one of strict reconnaissance, I did not stop for long. I drove through the slumbering township, then back along the motor road along the northern shore, heading for Inverness. There were many miles to go and trees obscured the view, so I drove quickly and it was not until reaching Urquhart Bay 15 miles to the eastward that I gained an uninterrupted view of the water; but here the road ran so high up on the mountain-side and the loch below was so very great in width, I knew it must be outside the range of my telephoto lens.

I drove into Inverness, and then back along the southern shore to Foyers, arriving just in time for breakfast. I ate a welcome meal of bacon and eggs, and felt much the better for it. The trip around the loch, 70 miles or so, though a waste of time in looking for the Monster, had taught me several lessons. In the first place in order to keep within camera range I would need to concentrate the search to the western half of the loch. Secondly, the canoe did not appear to offer any real advantage and might therefore be discarded. Thirdly, due to the heavy growth of the trees barring the view from two-thirds of the roads surrounding the loch it would be better to look for camera sites either at water level or above it on the hillside. As the latter seemed a more favourable proposition I set off to explore the hills to the west of the village of Foyers, driving the car as far as I could take it, then setting out on foot carrying my equipment.

I climbed to a point 1,000 feet above the water and then looking round decided to go no higher. To right and left as far as the eye could see the great loch stretched in panoramic splendour, and I realized at once that this was the ideal situation. With a really long-focus lens it would be possible to reduce the odds enormously from such a place— but without one, the extra height was merely a disadvantage. The lens I possessed was not of particular power, so I climbed down again to a lower level and spent the day sitting on a mossy ledge gazing down at the world around me, very much at ease. The sun shone, bringing with it warmth and colour, and a moving patchwork of light and shade on the surrounding hills and mountains, and as the day wore on I became attuned to the peaceful scene. Although there was no visible sign or disturbance in the still waters beneath, I felt the day had not been wasted, and returned to the hotel, sunburnt and cheerful.

After supper, standing on the patch of lawn outside the hotel, periodically scanning the loch with binoculars, I talked to the proprietor about the Monster. By now it was obvious what my intentions were and I had no desire or reason to conceal them. He had not been at the hotel for very long, but told me, with obvious sincerity, of a curious experience which he and his wife and two friends had had the previous spring, when watching the loch from the lawn on which we were now standing. Off the river mouth at Foyers, 600–700 yards distance, a large triangular object had appeared, sticking several feet out of the water; and this had suddenly shot off at incredible speed, travelling several hundred yards before submerging!

Mr. Rowand wisely preferred not to make any predictions as to what the object might have actually been, but did admit quite freely that it was very extraordinary and that the speed at which it moved reminded him of a naval torpedo, streaking through the water.

Later, I was to find out that he had at one time been an aeronautical engineer like myself, and the manner in which he told the story was fittingly reserved and specific, and yet carried conviction with it. I was at once conscious of the truly great difference between the words of a written account, and those of an actual witness.

I enquired whether he knew of any other witnesses with whom I might talk, and was told that Hugh Gray still lived in Foyers village, down below—the man who had first photographed the Monster in 1933.

I drove off down the hill at once, and found his cottage without difficulty but as Mr. Gray was out, arranged to call again the following evening. There were several hours of daylight left, and I spent these on Foyers Point, gradually increasing lens apertures as the light slowly diminished until I knew it would no longer be possible to film the Monster even if I saw it—and so ended the first day of the hunt. A day starting with near defeat, but ending in encouragement.

Second day

Daily log, April 19th 1960

4.45 am	Got up
5.20 am–9.10 am	Watched Borlum Bay–patrolled area 2
9.30 am–4.30 pm	Watched Foyers Bay
4.30 pm–5.15 pm	Interview with Mr. Gray
7.30 pm–9.30 pm	Surveyed and watched from area 3

RESULTS: Monster–nil. Survey–limited work. Interviews–one good. Total watching time 15.05 hours.

By now I had become thoroughly engrossed with the hunt; and once again I set off for Fort Augustus at dawn, and spent an hour or two watching Borlum Bay, perched high up on the mountainside.

The scene below was completely peaceful, the loch like a sheet of polished glass; and I stayed in position until the first sounds of life could be heard. The sudden bark of a dog, the clatter of milk bottles, and then a distant shouting—faint sounds, reaching up to my lofty perch, breaking the spell.

I coasted down the hill, and then patrolled the shore out towards Invermoriston, watching from the various selected places. Away from the road, and through the protected screen of trees, the far shore could be clearly seen and I searched every inch of it with binoculars, every rock and boulder, but without result, and then it began to rain. I retreated inside the car, and set up the camera on a shortened tripod in the place where the left-hand seat had been—an excellent arrangement, by which I could maintain the watch in any kind of weather. The telephoto lens peered out through the window, and by turning quickly in my seat I could bring the camera into action in a matter of seconds only.

Returning for breakfast, I continued to watch for the rest of the day in the manner recorded in the log, and at 4.30 pm as arranged met Mr. Gray in Foyers village. He proved to be a most courteous individual, and walked with me for a mile along the shore of the loch, to the exact spot where he claimed to have seen the Monster. He described his experience candidly, speaking with conviction; his account fitted the facts recorded by Mrs. Whyte, in every detail. Walking back to the village, he told me he had seen the wake the Monster made on several occasions, and described in graphic terms the extraordinary bow wave building up, rushing down the loch at remarkable speed, without anything visible making it. Back at his cottage he showed me a print from the original film he had taken, and although it was of poor quality it was possible to make out a sinuous shape in the water, and the ripples surrounding it, giving a fair impression of scale. I thanked Mr. Gray for the time and trouble he had taken, and drove off to continue the watch.

It was clear enough that people who had actually seen the Monster, or what they presumed to be the Monster, had no doubts about the hallucination, or being otherwise mistaken or misled. Because of this important fact I decided to try to talk to other witnesses—particular witnesses, people who by virtue of their work or profession were qualified to speak about the loch with assurance or authority. Alex Campbell, the water bailiff, for example, the man whose account I had first read in the magazine article, and the monks at St Benedict's Abbey facing directly on to the water. I knew I would get the truth from them. It would clearly be worthwhile sacrificing a few hours of watching time if I could talk to other witnesses as assured as Mr. Gray.

Well satisfied with the way things were going, I spent the rest of the day watching the loch from the mountainous terrain opposite Urquhart Castle, but without seeing any sign or monstrous disturbance.

Third day

Daily log, Wednesday, April 20th 1960

5.00 am	Got up, disgruntled
5.20 am–9.10 am	Watched Borlum Bay, and from area 2
9.30 am–10.30 am	Watched Foyers Bay

10.30 am–12.30 pm   Watched from "the Wall"
12.30 pm–2.30 pm   Visited Mrs. Constance Whyte
2.30 pm–6.00 pm   Watched from various places; northern shore
6.30 pm–8.00 pm   Watched Foyers Bay (good news)
8.30 pm–10.15 pm   Visited Fort Augustus. Talked to Father
                    A.J. Carruth at St Benedict's Abbey
RESULTS: Monster–possibility of V wake. Survey–continued.
Interviews–1 possible eyewitness. 2 non-witnesses. Total watching
time 10.50 hours.

I awoke with the alarm and journeyed to Fort Augustus, repeating the previous day's dawn activities. At the hotel the previous evening I had heard a curious rumour—a story circulating around the loch about a man who claimed to have seen the Monster partly out if the water on the shore, near a place called the "horse-shoe," a patch of scree, marking a precipitous slope on the inaccessible southern shore. The event was reported to have taken place seven or eight weeks previously, in February, and the man was said to have watched the creature for several minutes through binoculars!

There seemed to be no means of checking the story, but I was determined to try to find out what truth there might be in it, and that morning I sat for several hours opposite the "horse-shoe" mark, peering through binoculars. The far shore at this point stood a little over a mile distant, but with a good pair of glasses it was possible to pick out boulders and rocks of only a foot or so in diameter, and I realized that if the man really had seen the Monster out of the water, and if it was only half as big as people generally reported, it must have been very clearly visible.

I sat, hopefully, amongst the rocks, with the early morning mist shrouding the mountain tops and the loch as smooth as a looking glass reflecting every detail of the trees and the cliffs surrounding it. I sat without moving, conscious of the strange perfection of the place and the almost uncanny stillness—and then, a sudden explosion shook the ground around me. A tremendous thunderclap of noise, which echoed back and forth between the walls of rock, in rolling peals of thunder; grumbling and muttering, far off into the distance.

Intrigued by this curious incident, so early in the morning, two things struck me most forcibly about it. Without doubt, this was "the din of blasting" which was said to have frightened the Monsters and brought them to the surface in the early 1930's with such remarkable frequency. Having now experienced the stunning effect of an explosion within the confines of this rocky place, and the echoes that follow on, I could well imagine there might be something in the theory. The shock must have been felt underwater for miles, and I wondered whether repeated charges such as these might be used in a deliberate future experiment—an attempt to bring the Monster to the surface, without any risk of harming it. The idea was certainly worth consideration, but while I pondered on it another thought occurred to me. If the Monster was still alive, and not an hallucination, the final proof of its existence, of its shape and form, would come as another explosion—the news of which would rumble round the world, and echo into history!

I returned to Foyers for breakfast, and later on that day drove through Inverness to the village of Clachnaharry, there to meet Mrs. Constance Whyte, the author of the book. I had phoned her the day before and had arranged to discuss the Monster, a subject about which she obviously knew a very great deal. We sat and talked for an hour or two, engrossed in conversation, the focal point of which was of such great interest to both of us, and in this we enjoyed a mutual understanding, the foundation of a friendship which was to prove a great assistance in my future work at the loch. I learned the names of several outstanding witnesses, people who claimed to have seen the animal clearly, in some cases at a distance of only a few yards; in addition I found out much that concerned the history of the place, and of many unsuccessful private attempts to solve the mystery. I would have liked to extend the visit but I had to get back to the loch; I took my leave knowing that we would later meet again.

One thing I had learned in particular from talking to Mrs. Whyte—her book was the product of years of research, and although she had never seen the Monster herself, she had met and talked to a hundred people or so who had; and she possessed no doubts at all about the creature's reality.

Back at the loch, I spent the afternoon watching from the northern shore, returning to Foyers for supper to find that two of the guests had seen a curious V wake, earlier in the day, moving down the middle of the loch in a westerly direction. Questioning them closely, I found that neither witness had thought too seriously about it at the time, but the wash they described seemed to have been of considerable dimensions without anything visible causing it.

I knew that in the spring there were large salmon in the loch, weighing 20–30 lb, but as a fisherman too, I also knew it would be virtually impossible for fish of this size to leave a wake visible at a range of nearly a thousand yards!

Much encouraged by this first dramatic indication I decided to redouble my efforts and remain completely alert, but in spite of the urgent need to maintain a watch on the water I still had a number of important visits to make.

I wanted to meet the Benedictine monks, who lived at Fort Augustus. At 8.30 pm that evening I knocked on the door at the entrance hall of St Benedict's Abbey, there to meet one of the residents, Father J.A. Carruth, a tall man of engaging personality.

I explained the purpose of my visit, and before many minutes had passed we were engaged in conversation, and I was soon to find the subject I had come to discuss was one of particular interest to Father Carruth, who has published a booklet about it (a copy of which he kindly gave to me). Although he had never seen the Monster himself he was firmly convinced of its existence. We discussed the subject intently and came to agree that if the matter was approached with an open mind and the evidence studied carefully—particularly evidence at first hand—there could be no reasonable doubt as to the existence of a very large and extraordinary creature living in the loch.

I left the Abbey late that evening, gratified to have found that the conclusions I had drawn were shared by a man as sincere and intelligent as Father Carruth.

Fourth day

Daily log, Thursday, April 21th 1960

5.10 am   Got up
5.40 am–9.10 am   Watched Borlum Bay, and from Invermoriston
9.30 am–11.00 am   Watched Foyers Bay
11.00 am–11.30 am Interview with Colonel Grant of Knockie
11.45 am–12.30 pm Watched from southern shore–(area 4)
1.00 pm–6.00 pm   Interview with Alex Campbell, Water Bailiff

8.10 pm—   Filmed Monster off Foyers River mouth
RESULTS: Monster–approx. 20 ft of 16mm film exposed at 840
yards. Survey–useful. Interviews–two eyewitnesses, excellent. Total
viewing time 11.45 hours.

In the past five days I have driven nearly 1,000 miles, and during the
sorties round the loch climbed most of the high places that looked
down upon it—and I knew that fatigue was now my enemy. But,
there was still much to be done and after the usual morning watch,
as arranged, I met another witness: Colonel Grant of Knockie, who
lived on his estate high up on the mountainside beyond the southern
shore of the loch.

I was courteously received and we talked briefly. Colonel Grant
told me that on one occasion in November 1951 he had seen a great
disturbance in the water about 150 yards off the shore at Inchnacardoch Bay, and then the back of some large animal appeared. It had
dived, swimming just beneath the surface for a hundred yards or so,
travelling at quite extraordinary speed, leaving a wash like that from
a speedboat, breaking on the shore behind. As a result of this experience, he no longer had any doubt about the presence of some very
strange and powerful animal in the loch, but again, understandably,
he preferred not to try to name it.

I thanked him for his assistance, and permission to watch the
loch from the precipitous shore on his property. As I climbed away
over the hilltops, carrying my equipment I thought about the interview. I had been much impressed by the Colonel's reserved account,
which was obviously factual.

Reaching the crest of a hill I set up the camera high above the water, the wind whipping and snatching at my clothing. Looking down
on the great loch below I knew that unless it gave up its secret soon
I must fail in my quest, because within 24 hours I would be making
unwilling preparations for the long and tedious journey home.

That afternoon I continued to watch from Foyers Bay, and then
travelled once more to Fort Augustus, to meet a man I felt sure would
speak with authority on the subject—Alex Campbell, the water bailiff, whose story I had read many months ago. I knew that he had
worked as bailiff on the loch for many years.

Arriving at Fort Augustus I crossed the bridge over the river

Oich, and turned down a side road toward the loch, and a few moments later knocked on the door of a cottage. The man who opened it spoke with the pleasantly distinct accent of the Highlander, and, inviting me inside, quickly put me at my ease.

Alex Campbell was quite unlike the bailiff I had imagined. He was a man in his fifties, of slight build and scholarly appearance, and as I had already explained the purpose of my visit the day before, on the telephone, he proceeded with his story without further delay or formality. Briefly he described again the occasion on which he had watched the animal floating on the surface, with its head and neck gracefully upraised above the water with a great length of body showing behind it, and how it had dived in a swirl when frightened by the noise of herring drifters. He spoke also of other occasions in which he had seen the creature. Once when he was out fishing from a boat, with a friend, the great humped back of the animal had risen slowly above the surface, only a few yards distance, and had then sunk again—much to their disquiet and astonishment. On another occasion, at a greater distance, he had seen two separate Monsters, rolling and splashing about on the surface, one of which clearly exhibited a pair of forward flippers.

He talked with reserve, and absolute sincerity, making no attempt to impress me or dramatize his account, and in so doing he was doubly impressive. Thus it was that in the quiet of this cottage by the loch, I knew I had met a man who spoke the truth. As he was the first person to whom I had talked who had actually seen the Monster's sinuous head and neck protruding above the water, I knew without any last tremor of doubt that the huge back could not be that of a whale, or a porpoise, or seal, or any other type of ordinary creature, that had somehow contrived to find a way into the loch.

I said goodbye to Alex Campbell, assured of the fact that I had found a friend on whom I could count for support in the future. I left his cottage at 7.30 pm, and on the picturesque journey back to Foyers had time to re-assess the situation. In the light of what I had just been told, I decided to extend the hunt by a further day. I knew the Monster was a creature of flesh and blood, but more important, I believed it to be still alive. The quite unexplainable wake in the loch the previous day suggested this strongly.

Exhilarated by the thought of this last minute reprieve from failure, I set up the camera on the hill behind the bay at Foyers and began to watch again, and for the thousandth time scanned from left to right with every conscious effort. The light began to fail, and then, quite suddenly, looking down towards the mouth of the river I thought I could see a violent disturbance—a churning ring of rough water, centring about what appeared to be two long black shadows, or shapes, rising and falling in the water!

Without hesitation I focused the camera and methodically exposed 20 feet or so of film: and then, as the disturbance did not subside I decided try and get a great deal closer. I drove quickly down the zigzag road to Foyers, past the aluminum works and across the grass of a football field. Jumping out I hurriedly stalked through the trees and bushes dotting the small peninsular of land leading out to the river mouth. Scarcely daring to breathe, with camera at the ready, I approached the very spot, expecting to meet the Monster face to face—but all I found was the smooth-flowing river and the windswept loch beyond it; the disturbance seemed to have completely disappeared.

I was not unduly surprised, through very much disappointed. It was clear that the Monster, in the time it had taken me to get to the peninsular had swum away: so after half an hour of hopeful waiting, in the rapidly fading light I returned to the hotel—and had a glass of beer to celebrate. I went to bed quite sure that the Monster, or part of it at least, was nicely in the bag!

Fifth day

Daily log, April 22nd 1960

6.00 am	Got up
6.30 am–9.20 am	Watched from area 3
9.30 am–2.30 pm	Watched from Foyers Bay. Set up and filmed scale posts at River mouth.
2.30 pm–4.30 pm	Watched from boat, with salmon fisherman
5.00 pm–9.00 pm	Watched from Foyers Bay

RESULT: Monster–nil. Survey–shoreline study. Total watching time, 13 hours.

Relieved of the burden of anticipation, and the intolerable thought of defeat, I enjoyed a refreshing night's sleep; awoke without the aid of the alarm at 6 o'clock and travelling out along the road in an easterly direction, I climbed high up on the precipitous southern shore opposite Urquhart Castle and watched the loch until breakfast time. It was a beautiful tranquil morning; the huge expanse of water lay spread out below like a giant's tablecloth, sparkling in the early sunlight, and as I sat gazing down upon it I was startled by the noise of the stones tumbling down the mountainside. I looked up to see a roe deer bounding across the face of it—a lovely graceful creature, poised and alert, sure-footed on the almost precipitous slope, a very part of the natural scene surrounding it. I watched it disappear into a distant stand of fir trees and then looked back to the loch, conscious of the perfection about me.

Later in the day at Foyers I mapped the mouth of the river, and with the help of a boat charted its depths, using a weighted line and an adjusted fishing float; setting up some posts of measured length in the mud, I climbed to the point from which I had first seen the disturbance and took a number of covering pictures. I knew that the film would be of little scientific value unless some finite reference to scale could be included in it to provide a means of comparison.

While these operations were in progress I was helped by two local fishermen, who also lent me their boat and muscular assistance with every goodwill. During my days at the loch I had found that the people who lived around its shores, the proud Highlanders, accepted my strange activities with no more than mild curiosity and on every occasion had offered me their friendship and advice. I liked them, and envied their unhurried lives, and the magnificent countryside in which they lived.

In the afternoon, at the kind invitation of an ardent salmon fisherman, a retired sea captain, I went out trolling in a boat, and although we caught no fish it gave me an ideal opportunity to study a part of the shores of the loch and to watch from the water level. From this excursion I learned further lessons. A boat provided a means of studying parts of the shore of the loch which were otherwise invisible, but actual water level the view over the surface, the detailed view, was very much restricted; and furthermore the movement of the boat itself prevented the use of a telephoto lens. All things considered, therefore, it was a very poor platform.

Sixth and last day of hunt

Daily log, April 23rd 1960

5.00 am	Got up
5.20 am–8.40 am	Borlum Bay, and patrolled area 2
9.00 am–9.04 am	Filmed Monster for approx 4 minutes at 1,300 yards increasing to 1,800 yards.
9.07 am–10.00 am	Watched from shore west of Foyers
10.00 am–12.00 am	Filmed supporting sequences of boat
12.30 pm	Lunch
1.00 pm	Depart southwards END OF HUNT

FINAL RESULTS:
Loch survey: 482 miles covered; LOCH PERIMETER STUDIED
Total watching time: 73.00 hours
Interviews with informed persons: 2
Monster: Approx, 50 ft 16mm film exposed on back of large animal opposite Foyers Bay. 23.4.60

At dawn on this last day of the hunt, I got up and repeated the usual pre-breakfast activities: watching first Borlum Bay from the heights above, and then from the northern shore opposite the horse-shoe mark—waiting in hope to see the Monster climb out of the loch in the manner the rumour had suggested—but without avail. As the hours passed, I began to think of breakfast—the delicious sound and smell of frying eggs and bacon plagued my imagination, and my very empty stomach. At last I could stand it no longer and a few minutes before the accustomed hour set off back to Foyers, driving through the mountains along the single track I had come to know so well; past the little Loch Tarff, and the turning to Knockie lodge, then up and down and around about in the switchback of turns and gradients; finally climbing the hill behind the bay at Foyers, at the top of which the loch is seen once more.

A little before I approached this point I thought about the camera lying cushioned on the back seat. I knew that on the way down to the hotel I must pass within sight of the loch for a period of about 20 or 30 seconds but although I knew the rules about maintaining a state of instant readiness when anywhere near the water, for a moment I was undecided. It seemed a lot of bother to mount the camera

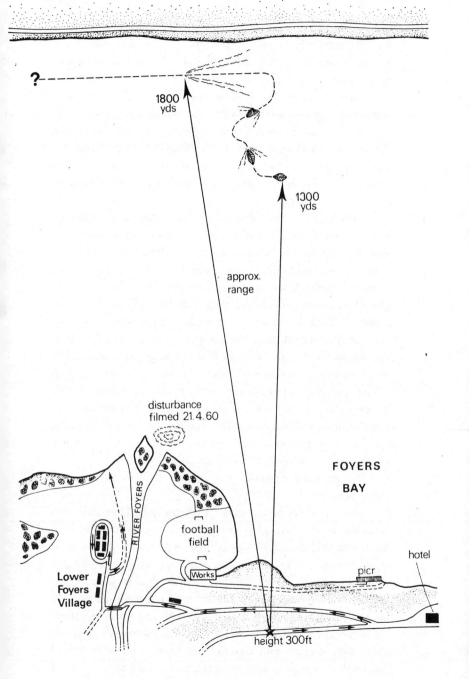

Sketch showing location of sighting, April 23, 1960.

and tripod inside the car again for just these fleeting seconds, and I wanted breakfast badly. After a pause, when everything hung in the balance, I decided to stick to the rigid drill which had become so much a matter of habit.

I stopped the car, and set up the tripod, next to the driver's seat, and putting up the camera adjusted the friction clamps for movement in pitch and traverse. I took a light reading, and adjusted the lens aperture, checking also the focus, the ciné camera speed and motor. When all was in order I trained the camera out of the window in a slightly downwards direction, repeating the actions I had come to know as if by instinct.

I rolled the car slowly down the hill, with one hand on the tripod, glancing down towards the loch, stretched out in panoramic view two hundred feet below. The far shore, though just over a mile away, looked near enough to touch, and the black water between lay without a ripple. At a point approximately halfway down the road to the hotel, looking out at the water, I saw an object on the surface about two-thirds of the way across the loch. By now, after so many hours of intensive watching, I was completely familiar with the effects that distance had on the scale of the local fishing boats, nearly all of which were built on common lines, 15 ft or so in length. The first thing that struck me immediately about the object was that although it appeared to be slightly shorter than a fishing boat, at the same distance, it stood too high out of the water; and furthermore, with the sun shining on it brightly it had a curious reddish brown hue about it which could be distinctly seen with the naked eye.

Unhurried, I stopped the car and raising my binoculars, focused them carefully.

The object was perfectly clear and now quite large. Although when first I had seen it, it lay sideways on, during the few seconds I'd taken with the binoculars it seemed to have turned away from me. It lay motionless on the water, a long oval shape, a distinct mahogany colour. On the left flank a huge dark blotch could be seen, like the dapple on a cow. For some reason it reminded me of the back of an African buffalo—it had fullness and girth and stood well above the water, and although I could see it from end to end there was no visible sign of a dorsal fin. And then. Abruptly. It began to move. I saw ripples break away from the further end, and I knew at once I was looking at the extraordinary humped back of some huge living creature!

I dropped my binoculars, and turned to the camera, and with deliberate and icy control, started to film; pressing the button, firing long and steady bursts of film like a machine gunner, stopping between to wind the clockwork motor. I could see the Monster through the optical camera sight (which enlarged slightly) making it appear very clear indeed; and as it swam away across the loch it changed course, leaving a glassy zigzag wake. And then it slowly began to submerge. At a point two or three hundred yards from the opposite shore, fully submerged, it turned abruptly left and proceeded parallel to it, throwing up a long V wash. It looked exactly like the tip of a submarine conning tower, just parting the surface, and as it proceeded westwards, I watched successive rhythmic bursts of foam break the surface—paddle strokes, with such a regular beat I instinctively started to count—one, two, three, four—pure white blobs of froth contrasting starkly against the black water surrounding, visible at 1,800 yards or so with the naked eye.

Awestruck, I filmed the beast as it proceeded westwards in a line as straight as an arrow, panning the camera to keep pace with it. I knew that as I had already exposed a length of film the day before there would not be much in reserve, and a quick look at the footage indicator proved this to be the case—I had only 15 ft remaining. Faced with the appalling decision, and only seconds to make it, I stopped filming. The Monster was now a long way off, and going at considerable speed in a westerly direction. I glanced at the second hand of my watch again—in 4 minutes the animal had swum nearly three-quarters of a mile, and was almost out of range: a mile and a half at least. I dare not risk these last few precious feet of film, because at any moment I knew it might come dashing back across the loch with head and neck upraised. It was the head and neck I wanted. I had now recorded the wake on 20–30 ft of film and could add nothing useful to it, so I decided on a sudden gamble—I knew it would be possible to drive the car across a field, right to the water's edge at a point to the west of lower Foyers and that in so doing in just a very few minutes, two or three at most, I could get nearly a thousand yards closer.

It was certainly worth the risk, and in seconds I folded up the tripod, and shot off down the steep zigzag road, going like a rocket, sounding the horn as I went, leaving a trail of dust. Over the bridge at the bottom, wheeling right I missed the entrance to the field, and cursing wildly carried on into a loop road round a group of houses, knowing it would prove the quickest way in which to double back.

I went round the tarmac circuit with tyres squealing, almost on two wheels, driving as I had never driven before in my life—and at the side of the road in front I saw a man look up, his face a mask of astonishment. Rounding the last bend and then down the track, I changed into a lower gear and tore off across the grass, arriving at the shore in a matter of seconds later.

I jumped out eager to learn my fate; but one brief glance was enough to tell me I had lost both the race, and my exhilarating gamble—the loch was once again as tranquil as a pond. Climbing 30 ft or so up a bank I looked to left and right, searching the surface with binoculars for miles in each direction but there was nothing to be seen upon it; no sight or even sound of a fishing boat, or other surface craft.

In the few minutes it had taken to race to the water's edge, the Monster had dived once more back into the depths, but before the dark waters closed over it, it had given up to me a part—a little part—of its quite uncanny secret; and although I now knew the hunt was really over, I also knew without any lingering shadow of doubt I had at last succeeded.

Through the magic lens of my camera I had reached out, across a thousand yards and more, to grasp the Monster by the tail.

BATTLE COMMENCES
Back at the hotel my news was well received by the other guests, most of whom, by now, had become infected with the virus of "Monster fever," which is very catching. At first I had been received with polite indifference: tolerance, in some cases, but there were certainly one or two who thought I was a little mad. But as the hunt progressed and people saw I was not, they began to ask questions when I returned for meals, at the end of each successive sortie, and it was because of these questions and genuine signs of interest that I was able to resolve a personal problem in connection with the Monster, and decided on a course of action—to which I have since adhered.

When arriving first at the loch, surrounded by cameras and equipment, I had been conscious of a feeling almost of guilt, and a strong desire to tell everyone I was only an ornithologist, on the look out for a rare species of birds—because I knew this to be a very usual excuse. It is quite extraordinary how many reports about the Monster have come from ardent "ornithologists" in the past, and it demonstrates a very natural human reaction; but I decided to tell people what I knew about the subject, everything about it, at every opportunity, and try to persuade them to study it themselves. In short, I started a sort of private crusade. In the past there had been so much running and hiding, and dishonesty around the loch, I knew it was time to start facing up to the facts, and the best way to do this was to start myself from the very beginning.

I did and from that moment on, if people asked me what on earth I was doing, I told them outright I was looking for the Loch Ness Monster—and I tried to tell them why, as well, if I had the time. It proved to be the right policy.

After breakfast I spoke with the proprietor about getting a boat to go out on the loch, so that I could film it as it steered the same course the Monster had taken. I knew that without a comparison of this kind the film would be of little scientific value, and yet with it the opposite would be true. The boat would provide a datum from which the Monster could be measured in terms of size, speed, and the sort of wake it made. I knew that this was absolutely necessary.

Hugh Rowand appreciated the need for this comparison at once, and without hesitation offered to do the job himself. This was generous, because I needed someone to steer and interpret the signals I intended to make with a flag, from the point where the film had been taken.

While Hugh prepared to launch the boat and install the 5-horse power motor, I scribbled out a briefing with a copy for myself, defining the course to take and the sequence of turns and actions. On it I made out the necessary list of visual signals, so that when we were separated by a mile or so of water we might still understand each other clearly. When everything was ready and we both understood the signal code, Hugh climbed into the boat and sat in the stern, canting the motor clear of the rocky bottom. I gave a shove, and it floated out stern first, and as I climbed back up the mountainside to the hotel I heard the motor splutter into life, and looking round, watched the boat slowly gathering way.

Back at the filming point I tied a white shirt to the end of a branch, in preparation, and watched fascinated as the 14-ft craft, now no bigger than a water beetle, made its way slowly across the loch leaving an elegant trail of ripples, fanning out on either side. The motor buzzed like a bumble-bee and drowned any possible chance of shouted words between us; and I was glad of the flag which retained our vital intelligence link. Hugh sat facing me and I could see the sunlight shining off his face, and the shiny aluminum motor cover.

It was a curious sensation, watching the boat crawl across the loch getting even smaller, and I began to realize the size and power of the animal I had seen. A couple of wags with the flag to the left put the boat back on course, and when it reached the approximate position where I had first seen the Monster I exposed a few feet of film. Then, as arranged, two or three hundred yards distance from the far shore the boat turned abruptly left and proceeded up the loch, following the course the animal had taken when just beneath the surface. I shot more film, and then waved another signal. The plan was working well. While the boat came back across the loch I opened the door of the car and took a flashlight picture, showing how the camera had been mounted inside, and then quickly filmed the boat again, stationary in the place where I had first seen the disturbance off the mouth of the Foyers River. Waving a last "come home" signal I jumped onto the car and made my way down the hill, turning right past an aluminum works at the bottom. As the boat coursed along at full throttle, sailing parallel to the shore, I paced it with the car, glancing at the speedometer. It was doing exactly seven miles an hour.

A few minutes later I was helping to heave the boat back out of the loch. Back at the hotel I sealed the camera with sticky paper, across which the Rowands signed, as witnesses. Half an hour later I said goodbye, and set off for Fort Augustus where I had one more call to make—the Post Office. There I concocted a lengthy cable which I sent to the directors of the British Museum, recording the fact that on 21 and 23 April 1960 I had exposed so many feet of film, through such and such a lens, with such and such camera, on the "phenomenon" known as the Loch Ness Monster, promising that in due course I would report on the result. I did this because the Museum represented the highest zoological authority in the country and I felt they ought to be the first people to know about it. This last duty fulfilled, I set off on the long journey southward, realizing I would not arrive at my brother's farm in Northumberland until two or three in the morning. I drove tensely without a glance at the mountain scenery sweeping by, conscious still of the same driving force, the compelling energy that had kept me on my feet through the days of the hunt. If the film was any good at all it would have to be shown to scientists in a manner that did not excite publicity, because I knew that the two did not mix well together. With the confusion of doubt and prejudice that swirled around the Monster, I realized that even a convincing strip of film would probably be received with chilling reserve by professional men, whose very reputations would depend on not being easily fooled. There was also the problem of my own status as an engineer, which did not qualify me to speak with authority on matters of zoology—and why should anyone take me, or my word for granted? For all they knew I might be a fraud, or a practical joker.

The more I thought about the problem, the more certain I became it was going to be a battle without any easy hopes of victory; and there was yet another problem which had to be anticipated. People already said the film would be worth a fortune, and although I was disinclined to believe them too readily there was always the possibility they might be right, and it be thus well to try and decide exactly what to do about it in advance.

If money had been the purpose of my visit to the loch, no doubt I would have been overjoyed at the thought of selling the film to the highest bidder; but as it was not, the prospect of a sudden flood of wealth proved almost an embarrassment—though a not unpleasant thought. Finally, however, I hit upon a compromise. I would not seek wealth for its own sake, and if I made anything out of the film, or from articles about it, the proceeds would go first of all towards proper equipment, and the cost of future expeditions. The cameras I possessed were inadequate, with the exception of the Bolex and 5 in. lens, which had been lent to me for the occasion. If I was to try my luck again at a future date I would need the very best equipment.

Thinking back to the episode of filming I realized the Monster had behaved itself extremely well, meeting the theoretical specification on almost every count. It had appeared in the early morning at a place where it was commonly seen—Foyers Bay, where, in point of fact, a Gaelic legend states explicitly that "water bulls are always to be found." It was the right shape and colour, a huge reddish brown triangular hump. At first when swimming slowly, it zigzagged about as if undecided where to go; then when just below the surface, it travelled as straight as an arrow at greater speed, disregarding the passage of traffic along the nearby motor road. Both these characteristics had been reported in the past.

Altogether it had performed in a most obliging manner, and although the film had been shot at extreme range, and the Monster had swum in the wrong direction and failed to expose its neck, I had no complaints against it. I had caught it by its tail, and no power on earth would make me let go of it.

By now I had reached the Ballachulish ferry, the quaint little boats splashing their way across the entrance to Loch Leven, and later drove through the towering haunted cleft of Glencoe, its chill winds whispering of black and evil deeds in centuries gone by—and then, by slow degree, out into the flat lands once more, into the pall of soot and smoke, amongst rows and rows of houses—the home and working place of very many people.

On Sunday evening, 24th April, I arrived home to enjoy a family welcome. I had already spoken to my wife about the film, and she was, of course, delighted. In her own way, she had contributed much to the success of the venture and understood its meaning and significance.

At the first opportunity I made contact with Kodak Limited, and arranged for the film to be processed and copied with every care and attention, and in this the works manager, Mr. Coppin, offered invaluable assistance. He witnessed the breaking of the seals and recording the film identification numbers, both on the original and copy films and, fittingly enough, shared with me the first projected viewing.

It proved to be a very tense business and as the film flickered on to the screen I watched it in dismay. The first few momentary sequences were badly under-exposed, showing boats and steamers moving through the water in different parts of the loch, which I had filmed deliberately, exposing a few frames on each to act as scaling markers. Then, running on for several minutes, the film portrayed the disturbance I had seen at the mouth of the Foyers River. At first sight it did look quite convincing, as though it might be caused by a powerful creature thrashing about underwater, but as the film ran on it became apparent it was no more than the wash and swirl of waves around a hidden shoal, caused by a sudden squall of wind. A second sequence taken the following day, with the loch in calmer mood, proved this to be the case without a doubt. Under the conditions of fading light and fatigue I had in fact been fooled completely.

By now the film had run for more than half its length—but then the picture changed, with dramatic suddenness. It threw into focus the strange humped back of the Monster, moving through the water exactly as I had watched it: slowly submerging, leaving a zigzag wake before turning abruptly left to proceed down the loch like a miniature submarine.

The film ran out, but we wound it back and ran it through again, watching carefully. It was on this second viewing I realized that although it recorded the Monster I had seen in essential detail it sadly lacked the colour, the contrast and perspective which had been so apparent at the loch. The shabby little black and white image that traced its way across the screen, though correct in shape and movement was indeed a poor imitation of what I had witnessed. In spite of that, it did provide the evidence needed, the proof of what I had seen myself in real life. More important still the two short filmed sequences of the boat steering a similar course provided the comparison of wash, scale and speed, so absolutely vital.

I left Kodak feeling cheerful enough, and in the weeks that followed showed the film to a number of scientific people, specialists of different kinds of senior standing, and also to a naturalist with whom I had corresponded. He was in fact one of the first people to see the film and, I think, appreciate its true significance and he did what he could to arouse official interest. It was not an easy task, and as time went by I realized the best I could hope to do at this early stage was to kindle people's interest sufficiently to start them on their own enquiring trail—nosing out the facts and stories and thus finding their own way to conviction. But, inevitably, it was a process that took time, and I had to admit my private disappointment. The one reaction I had not expected to meet was that of apathy, because the film demonstrated something of quite extraordinary interest—a huge animate object moving about in a freshwater lake cut off from the sea, in which there could, in theory, be nothing larger than salmon; a creature with a strange hump on its back, without a fin, capable of producing a wash greater than that of a Greenland whale!

This, I felt should be more than sufficient to prompt an immediate on-the-spot enquiry, with adequate equipment—but there were no signs of any such investigation or the funds to make one possible. For a time it baffled me completely, and I began to suspect that short of serving the Monster up on a dinner plate with a sprig of parsley on it, there was little one could do to get officialdom to act; but on further consideration, I decided not to be impatient. The majority of people with whom I was dealing were concerned, in one way or another, with a long established science—that of zoology—in which for very many years past the natural order of things and the animal life of this planet had been both well studied and understood, and in which discoveries were today measured in terms of minor variations of species which were already known. As long ago as 1812, Baron Georges Cuvier had stated with authority, "there is little hope of discovering new species of large quadrupeds!"

The fact that from then on a long succession of major discoveries proved the existence of some of the largest quadrupeds ever found on earth, did not mean that people anticipated any such discoveries today. Over the last 50 years only 3 or at most 4 large four-legged animals had been discovered, and to expect any sudden departure from this established line of thinking would be entirely mistaken.

But people seemed to be very complacent about the Monster. If it was exciting to me, an engineer, surely it ought to be ten times as exciting to zoologists—but the reactions I had so far encountered were so very staid that I began to wonder whether I might not profitably introduce the film in future with a stick of dynamite—like the one I had heard go off on the shores of the loch with such a delightful ear-splitting bang!

By now it was early June, and the film had remained a secret for nearly seven weeks. The dynamic reaction I had expected did not seem to be forthcoming, and in contrast to my effort at the loch, with such a specific object in view. I now seemed doomed to wait about hoping for something to happen; and worse, as I had now been asked to keep the film a secret indefinitely until a closer sequence was obtained, the opportunity of raising funds to cover expenses and the cost of new equipment had disappeared.

The expenses of the trip had turned out to be heavy, and I could ill afford them; I had underestimated the travelling costs, and by the time I had paid all the bills I was somewhat out of pocket. All things considered, the programme was not going well, and although I agreed to keep the film a secret indefinitely, it was a miserable decision to have to make. I was really rather proud of it.

Then, a newspaper reporter came to the door claiming he knew all about the film and the names of those connected with it. He made it abundantly clear that unless I told him the story he would very soon find someone who would! The situation was now entirely different and I wondered how to face it. If I continued to keep silent the story would come out anyway, no doubt distorted. The Monster would probably become a laughing stock again, but after so much effort this could not be allowed to happen. There was only one thing to do. If it was no longer possible to keep the film a secret I had to do the opposite—to show it to everyone at the first opportunity, and tell the truth about it; and this I resolved to do. After weeks of indecisive action it was nice to grasp the reins in my own hands once more, certain of direction; and besides, I was sick of all the secrecy, and the more I thought about it, the more certain I became it was the root of the trouble at Loch Ness. Over the past 28 years there had been so much secrecy, and sitting on the fence and dodging the issues that for all its sophistication and twentieth-century knowledge, science had in fact behaved towards the Monster, the "impossible," exactly as in Galileo's day—it had turned its back upon it. Such being the case the sooner people began to stick up for the truth about it, the better it would be.

As soon as it was possible I got on to an established firm of film distributors in London, and within hours a programme had been planned.

Arrangements were put in hand for an interview, and commentary on the film for release on television news around the world—an appointment was fixed with the director of a national newspaper; contact was established with the *Panorama* people on BBC television, and a written account and stills from the film were prepared for release to newspapers in other countries.

I had asked for action, and had got it, in no small measure. For this I had to thank two dynamic and honest individuals at the offices of United Press International, two Americans, full of fun and ideas; an absolute tonic to morale!

I had guessed that the greatest potential threat lay in mistaken or exaggerated newspaper stories. There was only one way to meet it, and that was to go right to the top, explain the situation and ask for co-operation. Several national newspapers had in the past treated the subject with consideration, and as the *Daily Mail* was no exception I had no doubts about offering them the story. As a result of the interview I was duly reported with good sense and impartiality on the morning of 13 June 1960. Reference was made to the *Panorama* show the same evening, so that readers had the opportunity to tune in and judge the film for themselves.

It was a relief to have taken the newspaper hurdle with such an easy stride, and once the story appeared in print it lost its exclusive value. This meant I could go forward to the next jump, the unnerving TV programme, knowing that the Monster's reputation was still undamaged.

The first interview with the *Panorama* team proved a success. They treated the matter with intelligence and understanding and although I could be sure the film and my own evidence would be treated with critical detachment, they would also give me every opportunity to prepare a case—and what was more important, they offered to enlarge the film as much as possible, and transpose it from the original 16-mm strip to 35-mm, a tricky laboratory process.

I was impressed with their efficiency and determination to adjudicate fairly, even to the extent of flying their questioner, Jim Mossman, up to the loch to make his own enquiries and satisfy himself I had taken the film from the place described—but however well the preparations were laid I could not hope to judge the outcome. This was to be the crucial test. The film would be judged by millions of people, and upon their reaction hung in balance the Monster's reputation; that delicate thing which had to be guarded so jealously! I knew well that if the film was not convincing, or if I failed to present my arguments properly, the results could be disastrous—but on the evening of 13 June I saw the enlarged version run through on the television screen for the first time, and was very much encouraged. Enlarged first two, and then four times, the humped back of the Monster could be clearly seen before it began to submerge. As it swam just underwater, close to the further shore, it was possible to make out a definite paddle action, swirling the water back in the manner of a breaststroke swimmer. But perhaps the most striking improvement on the original film (which was by no means spectacular) was the increase in definition and contrast resulting from a characteristic inherent in TV cameras, which produced a contrasted image on the screen from a low-contrast original. As the film was somewhat underexposed, and therefore short of contrast, it suited television well.

I was delighted to see that Alex Campbell had been invited down, and had no doubt he would speak with the same imperturbable manner I found so convincing when I met him at the loch.

The programme was, as usual, in three parts with the 'Monster' interview last of all, which meant we had to sit and wait in anxious anticipation. I did not enjoy this at all, and the confusion of lights and weird TV cameras, creeping about on rubber wheels, reaching out and up and down, goggling at us with a multitude of eyes, did nothing to ease the tension, and I was glad when the waiting time was over. The 12 minutes of question and answer, the film, and discussion of what the object could be, passed quickly and smoothly without embarrassment. Campbell spoke of his own experiences, of his view of the head and neck and the astounding speed of which the Monster was capable. I did my best to analyse the film step by step, pointing out the obvious difference between the wake of the boat, with its pronounced bow wave and line of propeller wash, and that of the Monster with its glassy V wash—but it was really unnecessary to talk, because the film spoke for itself, and I concentrated instead on a number of other points.

The film, I argued, clearly demonstrated the presence of some large animate object in Loch Ness, which in view of all the other evidence should now be properly investigated. I made no attempt to try to name the object because I knew that as I did not speak with authority on matters of zoology, it would be a mistake to try. Instead I showed a model of a Monster, made out of clay and carrying the same peculiar marks on its flank which I had seen so distinctly. The model was no more than a prototype, based on average statistics, showing the animal with three humps instead of the two most commonly reported, or the single hump that appeared in the film. A few days previously when building it, I had thought carefully about these humps and decided to include them only as an indication of the fact that they were seen on occasions. As future events were to prove, I might equally well have left them out.

The programme ran to its conclusion, timed to the second. When it was over Richard Dimbleby spoke to me and said that the film had convinced him the object we had watched was not a boat, or anything else he could think of—in short that it was quite unexplainable. This was good news, because Dimbleby was a man of very wide experience, a professional adjudicator used to summing up the pros and cons of controversial subjects. I felt this definite assurance spoke well for the film and its chance of success, and in this I was not mistaken. Within a few days mail began to come in from all parts of the country, from people in every walk of life: Lords and commoners, and from amongst all these letters there was but one dissenting voice—a man who rather gamely offered to explain the Monster by a meteorological theory.

There could be no doubt about it, the film had proved convincing and the Monster was firmly reinstated as an object of serious interest. I was happy about it; after so much disappointment we had gained ground, and having now done my best with the film, I could concentrate once more on my own particular studies. Having established the Monster was real and alive, I had got to stage three in the plan of investigation—the "what the devil is it?" stage, which promised to be intriguing.

By now I had the name of the man who claimed to have seen the beast on the shore, earlier in the year, and those of other people who had seen it at a distance of only a few yards with its head and neck out of the water. If I wanted to study the Monster in further detail the first thing would be to talk to these people. It would mean another expedition.

Chapter 7

The Hunt is On

The die was cast. What started out as an interest, a hobby, something to do with his mind during the long commutes to and from work had now taken on a life of its own and it all added up to a stark new reality. This was worldwide news. Like it or lump it, from June 13, 1960, Tim, and the Dinsdale name, would be forever linked with this legendary subject.

Whatever it was Tim filmed on that April morning was very much concurrent with his research of the previous twelve months. Sightings of a strange and unusual creature in Loch Ness had been going on for centuries; indeed, reports date back as far as AD 565 when St. Columba, the Abbot of Iona, was said to have saved a man swimming across the River Ness from a pursuing monster by invoking the word of Christ and sending the creature back to the depths, allowing the man to swim safely to the shore.

Many of the stories and legends of a large aquatic animal inhabiting the loch waters were mixed with Highland folklore and tales of the water kelpie; but Tim's research turned up consistencies between sightings, which made the subject all that much more compelling. Ordinary folks, from all walks of life, had experiences with Nessie. Reports ranged from what was at first thought to be an upturned boat sitting stationary in the middle of the loch suddenly bursting into life, moving across the surface at great speed, creating a bow wake as large as a speed boat before submerging out of sight, to those of a massive creature half out of the water with a long neck attached to a large strong body with diamond shaped flippers. Alex Campbell, the water bailiff at the loch for forty-plus years, claimed to have seen the beastie on a number of occasions. One of his sightings reads:

> It was a June morning in 1934. Then as the mist shredded away under the warm sunlight, I witnessed the most incredible sight I'd seen in my forty years as a water bailiff on Scotland's biggest loch. Something rose from the water like a monster of prehistoric times, measuring a full thirty feet from tip to tail. It had a long and sinuous neck and a flat reptilian head. Its skin was greyish black, tough looking, and just behind, where the neck joined the body, was a giant hump like that on a camel, though many times bigger. I pinched myself hard, but it was no dream, the Loch Ness Monster out there in the water was real and tangible…

Reports like these, and many others, fitted in with Tim's own extraordinary experience. As he mentioned, "the Monster behaved itself extremely well." In other words the characteristics shown by the monster during his encounter were consistent with so many of the descriptions from witnesses he'd either interviewed or whose reports he had studied. But now Tim's film showed the monster doing almost exactly what many of these people had said, which, certainly for Tim, just added to the authenticity of the whole subject.

Following Nessie's first public outing on the BBC *Panorama* programme, things started to happen. Tim was understandably hooked; so, while juggling family life with both work commitments and his growing passion for finding the truth about Loch Ness, a second trip to the loch was organized. Expedition #2 took place a few weeks after the TV show, with Tim returning to the loch in July. The aim: quite simply to improve on the film he'd just taken and so provide unequivocal proof of Nessie's existence. However, as Tim was to find out over the coming years, his "luck" of the first expedition was not going to be so readily repeated.

All sorts of folk started to contact Tim, many with their own Nessie close encounter; the report book started to fill, as did the character file. From this new flow of correspondence, primarily due to the exposure of the film on such a large media scale, reports came to light that were more and more intriguing, and his fascination—if it needed any encouragement—just grew. Tim followed up his two 1960 expeditions with a further two the following year. The results were disappointing. Nessie was starting to prove her reputation for shyness was not unfounded.

These periodic monster-hunting forays were starting to take their toll on both Tim's career and his pocket, as trips north were all self funded. Career-wise he was treading a fine line between being mentally engaged with his work and plotting his next rendezvous with the monster. As it was, the aircraft industry in the UK was going through a downturn, and so, much to Tim's delight—and Wendy's horror—he was made redundant. Now he had all the time in the world to concentrate on snaring Nessie. There was just the small irritation of how to pay for it and feed a family.

This was a very serious concern; the four trips to Scotland had depleted the family's account substantially. He found himself unemployed yet driven toward the goal of proving the monster's existence, a pursuit which so far had only cost money. Wendy, not for the first (nor last) time, solved the problem. Her sharp eyes spotted an advert in the local newspaper for a life insurance salesman. Although Tim had no experience with insurance he did wholeheartedly believe in the product due to the fact that his own mother's insurance policy was paying out after his father's death; if she hadn't had such a security net she would have been left destitute, as the family's wealth had been destroyed during the war. Tim secured the job. The most appealing part of it was that basically he would be self-employed, meaning he had the freedom to take prolonged absences from work to continue his research and go on further monster hunting trips. It was a very agreeable solution and one that Wendy found particularly pleasing. With four children under the age of eight to look after she was happy to have her husband out of the house making a living and not getting under her feet all day.

With the burden of providing an income somewhat taken care of, Tim could get down to the serious business of chasing Nessie.

Highland legend depicts "there are always water bulls to be found at Foyers," and having filmed the animal from the road overlooking the stretch of the loch known as Foyers Bay (named after the tiny village built up around the river running into the loch), Tim felt it was an obvious place to concentrate his efforts. His aim was to get close-up, undisputable photography of the monster. To achieve this, he knew he needed to get closer to the loch-side. In 1962 he found what, on first sight, seemed an ideal spot just along the shoreline from the mouth of the

Foyers River. It's not a particularly large watercourse, but one that flows from high up in the mountains behind the village, working its way down over the rocks and waterfalls to eventually merge with the loch.

I would refer to it as 'the place marked X', because it was not shown on the inch-to-the-mile Ordnance Survey map, and needed an X to point it out.

It was situated to the east of Foyers Bay on a strip of beautiful, inaccessible shoreline. To get there involved climbing 150 feet up the steeply wooded shore, beneath the trunk of a fallen tree, steeply, to a tiny shingle beach, perhaps twenty yards in width, with an old tumbledown wooden boatshed set up against the mountainside, so overgrown it was like a cave in appearance.

Completely secluded, it made a natural hide where I could set up my cameras and equipment in ambush, and during the 480 hours I spent watching from there when I needed food I could leave the equipment behind unguarded. And yet, despite these advantages there was something odd about 'the place marked X.'

Four of the five expeditions based there resulted in illness or injury, or accident of a most unpleasant nature, and I became aware of some strange influence which seemed to be malevolent and prompted me to examine the background history of the area, and to research the water kelpie legend in particular…

The water kelpie is part of Highland legend. Folklore says the water kelpie, or water horse, appears in the form of a handsome stallion that lives at the bottom of fast flowing rivers and streams. When a traveler wants to cross, the water kelpie emerges to offer a safe ride to the other side. Once the traveler has mounted, the water kelpie's skin becomes adhesive, trapping the rider who is carried off to the bottom of the river to be drowned and devoured.

Foyers and the surrounding hills had indeed seen a fair share of nastiness over the centuries. Feuding clans hacking each other to pieces, while not commonplace, certainly occurred, and the battlefield of one such skirmish was in close proximity to "the place marked X." This and some of the other stories of unpleasant acts and violent deeds known to have taken place in the area may well have added to his feelings of disquiet.

Whenever Tim talked of "the place marked X" it was always tainted with unease. He put in a lot of hours shore watching from the spot, but never felt comfortable there and certainly felt no affection for those first expedition years. He was cursed with bad luck on every one of those trips, beset with sickness, appalling weather, equipment failure, and even once being caught in a sudden squall while paddling a small boat along the shoreline and almost capsizing. All of these incidents contributed to his growing dislike for the site. Whether it was the secluded position—which was the initial attraction—or something else, he didn't know. Between 1962 and 64 Tim made five miserable trips to the place marked X without so much as a sniff of Nessie, so he decided it was time to review his watching strategy.

While the BBC *Panorama* programme highlighting Tim's film had, for the most part, got people's attention, the scientific establishment continued to be distant, reserving opinion. And, as Tim once put it, they were probably waiting, wanting Nessie served up on a silver plate with a sprig of parsley before they would come out and openly agree there was something more in the loch than science currently accepted.

Others had taken action. During 1961, Scotsman David James MP, MBE, DSC, took up the baton and led the charge to develop a non-profit company to take on the challenge of investigating Loch Ness. The company's first incarnation was the Loch Ness Phenomena Investigation Bureau. This title was quite a mouthful and so was quickly shortened by dropping the word "phenomena." "The LNIB," as it would become known, was eventually shortened further still to just "the LNI." This was the acronym used by those who worked as volunteers or were in any way involved with the organization, like the independent monster hunters.

The LNIB started life in 1962 with the aim of collecting sighting information and general Nessie evidence. It performed fieldwork and set up groups of volunteers to do a variety of watching and research while putting a respectable face on monster hunting.

David James and the board of directors ran the LNIB for the next ten years, and hundreds of people volunteered their time, their summer vacations, and their holidays to come and watch the loch. As the years passed, the organisation developed into a well-run, serious investigating unit. David James, or DJ as he became known, was instrumental in gaining sponsorships for much-needed equipment. He was responsible for the appearance of 35mm ciné cameras with large telephoto lenses. The cameras became the LNIB trademark and gave hunting Nessie a new dimension. Although Tim's film was fascinating, making the viewer ask, "What the devil is it?" the overall film quality wasn't all that great. It was slightly underexposed and the target was at such extreme range that the film didn't give the full impression of the monster as Tim had seen it when viewed through his binoculars. The acquisition of such top quality, state-of-art, photographic equipment put the chances of gaining clear and undisputable pictorial evidence on a much higher footing.

From the humble, yet respectable, beginnings the summer expeditions slowly grew. The original two weeks of operational hunts expanded into months and then, eventually, the whole summer. May to October would be taken up with volunteers coming for, mostly, a fortnight's stint doing a variety of tasks which all amounted to monster hunting.

After a couple of years of nomadic-style shore watching, in 1965 the LNIB moved to what would become their permanent home. A farmer's field on the north shore, about a mile or two west of Urquhart Castle in an area known as Achnahannet, provided an ideal, almost central, position for their headquarters. Situated on the roadside about 200 feet above the surface of the loch, the site commanded an impressive view for miles in both directions. With the addition of the new long lenses the camera's eye would be within range if Nessie decided to surface in the vicinity.

The Achnahannet site started life with a couple of old caravans where the crews would sleep. These were eventually expanded with the addition of a kitchen and mess hall, a workshop, and viewing/camera platform. The caravans were all painted green and arranged in a uniform manner giving the site an air of professionalism. Before long, the place became the hub for all things connected to hunting Nessie.

The volunteers would arrive for their two-week stint and be split into crews for the duration of their stay. The crews had their daily tasks. If they were on watch then they would be designated a watching site. This would entail taking one of the expedition vans (some of which had been modified with the addition of a camera platform fixed to the roof making it a mobile filming unit) to their designated site. Once there, they would clamber onto the top of the van, fix a camera to the pre-positioned tripod, and then sit and watch the loch. Shore watching could be mind-numbingly boring, so it didn't take long before individuals found their favourite sites. Strone, overlooking Urquhart Bay, was least popular due to being on the top of a garage at the end of a long private lane, which meant for a lonely day of watching due to the lack of passing tourists. Dores was the furthest away from HQ, so entailed a long drive, and again, situated in a quiet roadside lay-by could make for a slow day of hunting. Foyers, on the other hand, was the firm favourite; close to the road and water's edge, it was a natural stopping point for tourists. Interested holidaymakers would readily pull in, and go over to chat to the person sitting on top of the distinctive LNIB van. The most commonly asked question was, "What time does Nessie show up?" In spite of annoyance at some of the more repetitive questions, it was generally agreed that Foyers was the best spot to spend the day due to the friendly tourists and, more often than not, genuine interest in the monster and the LNIB. There was also the added bonus of generous offerings of tea and sandwiches.

Those not on mobile watch would be employed around HQ, where tasks and duties abounded: helping in the kitchen, cleaning, staffing the small information centre, building a harbour, and lending a hand with a variety of ongoing experiments; plus HQ had its own camera site, so there was also watch duty. Over the years, Achnahannet became an exciting place to be and a hive of monster-hunting activity.

In 1964, after lonely years of shore watching from "the place marked X," Tim felt the need for a change and decided to join forces with the LNIB. Their site—pre-Achnahannet days—was perched on the battlements of Urquhart Castle, overlooking the bay where the 1955 MacNab[1] picture was taken. It was a welcome change from the solitude of those early trips and coincided with the arrival of the volunteer monster hunters. The participation of these volunteers became a tradition and a culture that carried through the following years of the search. In contrast to his previous solo experiences, the two expeditions he spent at the castle were enjoyable times, even if they were lacking in results and the weather was appalling. Nevertheless, although the experience of watching with others was a pleasant interlude, the allure of getting back to being an independent monster hunter was strong. Tim felt there was unfinished business at Foyers; after all it was the area

1 In July 1955, bank manager Peter MacNab took a photo of what appears to be the back of a large animal swimming toward Urquhart Castle. Using the sixty-four-foot castle wall as a measurement, the animal appears to be greater than forty feet in length.

where he'd filmed the monster and where there had been numerous accounts of sightings going back over the years. He just needed to find a new base camp. His old spot, "the place marked X," was no more. Since his last expedition there in 1963, the shoreline along that area had been completely destroyed to make way for the building of the Foyers hydro works. Tim had no intention of returning to his old viewing spot, but needed a place equally as private to replace it.

Loch Ness is basically a massive trench, part of a fault in the earth's crust which runs from the east to west coast of Scotland, and which makes up The Great Glen. The surrounding mountains have been carved smooth by the last ice age and the sides of the loch are steep, with the majority of the shoreline being a continuation of the mountains going down vertically many hundreds of feet straight to the loch bed. Occasionally there are areas where the shore juts out a few metres into the loch, creating a gradual slope, but then all of a sudden it comes to an edge and plummets into the chasm-like depths. So when Tim discovered a small treed area right at the mouth of the Foyers River—actually a tiny island—he was delighted. It offered the privacy he craved yet was close enough to humanity to avoid the lonely solitude he had endured during his earlier expeditions. The water view was unrestricted and with ample space to pitch a tent. All in all, the island, as it was to become known, was a very agreeable solution for his expedition needs.

It was about this time I remember becoming aware of the monster. As a five-year-old I was to develop a tradition that would last for many years. The day before Dad was to leave on expedition, I would insist on being taken to the local green grocer store where I'd pick out the biggest carrot I could find so Dad could feed it to Nessie. It was just a childlike thing to do, but for me it was very important. My dad was off to hunt monsters—albeit friendly monsters—and so in my eyes he needed something to give her when she came up to the surface to have her picture taken; a bit like a treat you'd give a household pet.

It was also at this time when I began to get a sense of being a part of something more than just Mum, Dad, and home. News of Sir Winston Churchill's death sent quite a sombre feeling through our household, with Dad saying we'd lost the country's greatest-ever statesman. Having spent much of his childhood in a colonial-style upbringing and having just lived through the Second World War, Dad had a strong sense of national pride. So when Sir Winston Churchill passed away in January

1965, the whole family sat in front of the black and white TV to watch the state funeral take place in London. The coffin was transported a short distance down the River Thames, passing the East End docks which had been so hard hit during the London Blitz. The Dockers lowered their crane jibs to salute the wartime leader as his body passed. The nineteen field-gun salute was followed by the coffin being loaded onto a special train at Waterloo station bound for Baldon, Oxfordshire, and his family's burial plot at St Mary's Church, close to his birthplace of Blenheim Palace.

The train route would pass very close to our house. Once Tim knew this, we were all squeezed into our green Austin Mini to go and find a spot where we could pay our respects. Not surprisingly, the two train stations we passed were completely full with folks who had the same idea, so we drove on to a nearby town called Pangbourne where, in places, the line was a little more accessible. We parked, precariously, at the side of the road and found an opening in the fence next to a bridge. We all clambered through and negotiated the steep bank leading to the rail lines to watch the train pass. We didn't have to wait long before seeing the grand old steam locomotive, itself named Winston Churchill, chugging its way toward us. There, for a fleeting moment I saw through the glass-sided carriage to the great man's coffin, a soldier standing at either end, head bowed, providing a guard of honour. I saluted as the train passed. My father said the experience would stay with us for a lifetime, and he was right—it has.

Tim had a healthy respect for the British establishment. He was proud to have been born in Wales of English parents (his father was of English parentage only himself born in Yokohama, Japan) and ply his trade in Scotland—a true member of the Great British nation. During the years of the search he would periodically contact the Duke of Edinburgh's office at Buckingham Palace. Tim had sent a copy of his first book, *Loch Ness Monster*, to Prince Phillip who, reportedly, enjoyed it and asked to be kept in the loop of happenings at the loch. The subsequent Palace correspondence was kept in a file in Nutter's Nook (see chapter 19) under the title of "Is & ER's."

Chapter 8

A Very Important Piece of Equipment

Strange things started to appear at our house. At first there was an ex-army, camouflaged bell tent, which was to be used on the forthcoming island trips. The tent was, of course, set up in the back garden, and when Dad wasn't looking it became the centre of entertainment for us kids. We would spend many hours creating adventures, chasing imaginary creatures and being chased back. It was all great, harmless fun, which just added to the excitement of Dad leaving to go and hunt monsters.

The next piece of equipment to arrive was a little more serious and far more impressive: a crossbow. This was a powerful weapon with a range of over 500 yards. During Tim's investigations he'd come across occasions when people had reported seeing the monster for more than a few moments; indeed there had been incidents when Nessie had been observed on the surface for many minutes at a time. With this in mind, Tim's plan was—if he were lucky enough to experience such a prolonged sighting—to use the crossbow to obtain a sample of Nessie's hide. It was a bold idea that would need a few kinks ironing out before being put into action. First, there was the problem of designing and developing a suitable arrow and head that could not only hit the monster, taking a skin sample while inflicting minimal injury, but was also retrievable. Tim had tried using a long bow, but the distance he could get was very short. The problem with an arrow that has a weighted tip is that unless it's released under great power it falls to the ground within twenty or so yards. The crossbow was in a different class; when it came to power and accuracy it left the long bow obsolete. However, the arrowheads still needed refining and over the coming years a number of prototypes were developed. Test firing took place mostly at the local archery range, but once in a while some arrows were shot off in the back garden. An old carpet would be draped over the washing line as the target and firing would commence. Not the safest of environments when our house backed on to another, and, of course, the inevitable happened: an arrow deflected off the carpet and up onto the neighbour's roof. It could have gone anywhere. Unperturbed, but with a little more sense of the danger of unleashing projectiles in a residential area, Tim returned to the archery range. With further refinement the arrows started to fly straighter and on a truer trajectory. The next problem was how to retrieve the arrow once it had made contact with Nessie. Attaching a large seagoing fishing reel to the handle of the crossbow theoretically solved this—but on the first firing the reel got tangled, stopping the arrow in mid flight returning it directly towards Tim at a similarly high velocity! It was a work in progress.

The crossbow was a piece of equipment we kids were only allowed to admire from a distance. My friends and I would, occasionally, be allowed to view it in its specially designed case. The case was very impressive, decked out in sky blue velvet with separate compartments for the brightly painted arrows, clips holding the arsenal of fierce looking skin-penetrating darts and the crossbow with a telescopic sight mounted atop—all things to make young boys' eyes go wide with delight. It was also to start the tradition of "this is a very important piece of equipment," meaning young fingers weren't allowed to touch. In fact, that pretty much became the mantra for all of the expedition equipment, and one which, as we grew up, we would tease Dad mercilessly about.

As Tim became more independent, the equipment count also grew: tent, cooking tools, sleeping bag, cameras, the crossbow, and then a nine-foot inflatable dinghy joined the ever expanding collection of the monster hunter's paraphernalia. The small Avon vessel would, over the years, become a much loved and treasured work horse, accompanying Tim on countless expeditions, and a number of family summer holidays too. It affectionately became known as the Moo-scow and will feature again, many times, later in our story.

The first island expedition was a success in that nothing untoward happened, with none of the difficulties of the earlier solo trips being repeated; but still Nessie didn't show. However, a tradition was born. Tim, living on the island in the ex-army bell tent, would only get to shave every few days, so he started to leave a small tuft of hair on his chin, which soon became long enough to be recognized as a sort of beard. It was quite distinctive and, at first, was always removed a short time after his return to Reading. But, as the years passed, Tim's "chin wig" as he liked to call it, became a feature, something he was recognised for, one of his many slightly eccentric, and yet endearing, characteristics.

The ongoing battle to find Nessie was starting to take its toll on both Tim and Wendy. Although Tim's resolve was to gain conclusive, indisputable evidence and thus prove the truth of the monster's existence, the price of achieving this both in monetary terms and family cost was turning out to be high. Wendy, while she fully supported Tim in his endeavours was, understandably, starting to be concerned about where it was all leading. Tim recognized if it wasn't for Wendy's staunch support it would be impossible to continue with his expeditions. The years were rolling by and after hundreds of hours of shore watching the results weren't all that exciting. With nothing tangible to show for all the effort, time and money dedicated toward the cause, the Dinsdales needed a shot in the arm of enthusiasm. It was to come not in the form of a new piece of evidence—although that would have been welcomed—but via the confirmation of Tim's own 1960 footage. In 1965, David James, through his contacts at the House of Commons, had arranged for a film sequence obtained by the LNIB to be submitted to the RAF for analysis. Impressed with the results, he suggested Tim do the same with his own film.

The RAF's Joint Air Reconnaissance Intelligence Centre (JARIC) duly took the job on. Tim was content they were the right body to undertake the task. For one thing, they were the recognized experts in this field in the country, and, for another, they would come by conclusions with a scientific impartiality. JARIC presented their finds in a report in January 1966. The results were massively encouraging. Tim wrote:

> The report accepted the film as genuine on the basis of pure mensuration [measurement] of the original filmed image, stated that the object seen was neither a surface boat nor a submarine—'which leaves the conclusion it is probably an animate object.' When first on the surface it stood 3ft high; allowing for the underwater bulk, 'a cross section through the object would be not less than 6 feet wide and 5

feet high.' At first there had been some confusion over estimates of length. No estimates were made in the report. Allowing for the niceties of photogrammetric work, and interpretation, and for certain criteria applying to the solid object filmed, which had both height and width, it was stated that the 'residual length in the horizontal plane would be in the order of 12 to 16 feet.' The speed of the object was estimated as between 7 and 10 mph as a mean...

This was a welcome vindication of the film's authenticity; it also gave rise to a second wave of interest in the whole subject. Of course the dissenters still existed, but now they had to argue against a reputable institution that had nothing to gain from its report.

The JARIC report bolstered Tim immensely; he had been lecturing sporadically on the subject for a number of years, but now with this new, official, confirmation the phone started to ring more regularly and with it the next expedition was planned.

Besides the excitement surrounding the analysis of the film, there were two other noteworthy happenings that year: Wendy started her career with a government-funded organization called The Citizens Advice Bureau (C.A.B.)[1]; the other was England winning the football (soccer) World Cup against their old foe Germany. I have fond memories of being packed off in the car with my mother and two sisters to a family friend's house so my father and brother Simon could watch the match "in peace." It was a grand afternoon of watching the old black and white telly, the house stuffed full of little kids all running around, over excited about the match. Later, after a nerve-racking game, and as the final whistle was about to be blown, Geoff Hurst scored an incredible winning goal. As the TV commentator muttered those immortal words, "they think it's all over—it is now!" my mother's friend, overcome with joy, let fly her three-month-old baby high in to the air in wild celebration. It was a happy time.

The winter months were the time for lecturing and planning next summer's hunt. Tim's reputation as a guest speaker was growing, as was the list of universities he entertained. Sheffield, Edinburgh, St. Andrews, Aberdeen, and Glasgow, along with numerable schools and colleges, were now all on his portfolio. Having no formal education in either

1 A nationwide government service developed during the Second World War to support individuals and families by providing free, confidential advice and assistance.

biology or zoology, Tim was quite aware of the honour he was being paid when invited to talk at these classic learning institutions. And the fees, although relatively small, all helped toward the ever-growing cost of chasing Nessie.

Armed with a renewed enthusiasm, the next round of experiments was conceived. Not for the last time the idea of dropping bait into the loch was tabled. This, the first try at tempting the monster to the surface, was to come in the form of asafetida[2] an incredibly pungent root extract taken from a plant in eastern Pakistan. A portion of it was dropped into the loch just off the edge of the island where Tim and fellow monster hunter John Addey waited with cameras at the ready for an excited Nessie to show. Needless to say their efforts went unrewarded, with Nessie being less than impressed with the overpoweringly smelling root, and the local shopkeeper being equally unimpressed with the two adventurers who reeked of the pungent odour when they went ashore to collect supplies. In fact, a year or so later, Tim chanced upon a report that Norwegian fishermen would use the same smelly concoction to ward *off* sea monsters.

Despite the lack of results from really whiffy experiments, the island became a well-liked, and favoured, watching spot. Being alone was becoming enjoyable; there's a connection with nature when one spends so much time living amongst it. The relative inaccessibility of the island to humans provided a haven for nesting birds, and oystercatchers in particular found the habitat very appealing, laying a cluster of speckled eggs in a small indentation in the stones just above the high water mark. For the most part, Tim lived in harmony with the birds and other small animals on the island; but this changed one day when, inadvertently, he stepped on a nest crushing the eggs. The hen was so incensed at his clumsiness she attacked his tent. Relations between Tim and the nesting oystercatchers never really returned to their carefree ways prior to the accident.

A year earlier, my dad had pinched an egg to give to my brother Simon for his collection. Although I wasn't the recipient, I was very proud of the egg, as none of my friends had one. It was quite a prize. Showing it off to my mates one day, I accidentally dropped it and

---

2   Referred to as the world's smelliest spice and also known as devil's dung, it is extruded from the root of species of *Ferula*, flowering plants found in the mountains of Afghanistan, Pakistan and India.

watched it smash on the concrete floor of our garage. Repairs were hastily made but the egg was never to be the same. Simon didn't seem all that worried, but I was devastated and I repeatedly asked for a replacement. But on each returning expedition I was told the birds had all finished laying their eggs, which had hatched by the time he arrived at the island. I always suspected that Dad felt guilty about destroying the birds' nest and didn't want to encroach on the oystercatchers' hospitality any further. A replacement egg never did materialize.

The equipment list was growing again. The acquisition of a twelve-foot wooden stepladder was one of the less exciting pieces, however, with a very practical use. Tim would spend hours sitting atop the ladder viewing the loch from the island's shingle beach, which gained him the occasional odd stare from passing fishermen; but the extra twelve feet in height increased his field of vision considerably. He must have cut a surreal figure sitting there perched on top of the ladder with a camera in one hand and binoculars strung around his neck while he scanned the loch.

The next purchase was far more exciting: a 1957 MK8 Jaguar. After years of being bundled into the family's Austin Mini, and making trips to the shops in Wendy's original Messerschmitt bubble car, the Jaguar seemed huge. Tim justified the new car as a must-have because of all the expedition equipment he needed to transport to and from the loch. It was truly a monstrous vehicle with grand sweeping lines, a two-tone blue paint job and a lithe looking jaguar on top of an imposing grille. Inside were two large, comfortable, red leather-clad bench seats and the dash looked quite stunning in a walnut veneer. I was so impressed with the rear passenger area as there were two pull-down wooden tables fixed in the back of the front seats. And, if that wasn't enough, the final crème-de-la-crème was the cigar lighters—with no less than three of them conveniently positioned throughout the interior. It truly was a car fit for Cruella de Vil.[3]

Nothing like it had ever been seen on our street before, and neighbours would stare and stop to ask questions. Dad would introduce it as the "expedition car" and explain its practical use. I, on the other hand, just wanted to play in it, and I spent hours making up games, imagining it as a fort to be defended at all costs or disappearing into the cavernous

3   The evil character from *One Hundred and One Dalmations*.

boot during a game of hide-and-seek. However, it wasn't all good; the Jag for all its finery on the outside was something of a problem child under the bonnet. In other words, there was a reason it only cost £40. A lot of time was dedicated to repairing and preparing it for journeys long and short. Breakdowns were commonplace and due to one such incident my sister Dawn refused ever to ride in it again. The car died right in the middle of the road causing a traffic jam in both directions. Dad was taking us to a school event, so all our friends were there to witness the breakdown. Dawn was just too embarrassed to jump out and help push it to the side of the road and sat there ducked down under the seat while Dad and I did all the heavy work. Of course Dawn was seen in the Jag again, but she stuck to her guns and never accepted another ride in it to school.

The temperamental Jaguar just added to the list of equipment with personalities; however the next piece of monster hunting kit to arrive was very special indeed. From Tim's first expedition he had either borrowed or hired cameras, the cost of which was prohibitive and ate yet further into the meagre budget Tim was able to allocate for his trips. But on this point he was stuck: without good photographic equipment there wasn't really much point in even going to the loch. Good cameras with long lenses were the tools of his trade and he had to have the best he could afford. The simple answer would have been to buy what he needed outright, however this equipment was very expensive and far out of his financial reach at the time. Book sales were good, as was the lecture circuit, but both of these collectively didn't bring in nearly enough to splash out on such an extravagance. It was a recurring dilemma with no end in sight.

The solution came quite literally out of the blue. The JARIC report of 1966 had rekindled the public attention to the subject and, with the hard work of both the independent monster hunters and the LNIB, there seemed now to be an appetite of public acceptance that the loch harboured something more than met the eye. The following year, Tim received a scientific award from the film and camera company Kodak. His prize was to purchase all the photographic equipment he needed to continue his research at Loch Ness. It was amazing. Both Tim and Wendy were stunned and delighted at this present from heaven landing in their lap. It added to the renewed enthusiasm and, if it was ever in doubt, the hunt was back on!

The weeks following the award saw numerous boxes and crates delivered to our house, all marked for Dad's attention. We waited with bated breath to see what the next exciting thing to be unwrapped was. Of course it was like Christmas, not only for us kids but Mum and Dad were also thrilled at the influx of such top quality equipment which, before the prize, would have only been a dream. A large, very professional-looking tripod was followed by a distinctive, oversized, black hard-backed carrying case for the camera and lenses. The inside was very impressive with every speck of space being accounted for; the modern Beaulieu 16mm ciné camera fitted perfectly with the distinctive shape of the twin film spools sliding snugly into position. The three lenses of varying sizes all had their designated spots, filters, batteries and spare film, everything had its place; it was a custom built case like none the Dinsdale family had ever seen before.

With this windfall came the inevitable "this is a very important piece of equipment." We truly weren't allowed to touch, hold, look at, nor investigate any of it. Indeed the closest I came to any part of it was to put my eye to the camera's viewfinder, but that was only after being given express permission by my father. Tim knew the value of it all, not only in monetary terms but also the value it added to his research and edge it would give him to gain that indisputable piece of evidence he had always held as his goal. He now had the perfect tools to do the job.

Once he'd familiarized himself with all the knobs and gadgets and generally got to know how the whole thing worked, he set about customizing the tripod to take a whole slew of cameras and sound devices. The months that followed saw Tim spending many hours in the garage attached to the side of the house working on the development of what would be christened the "Cyclops Rig." Tim designed and built an ingenious contraption where the new Beaulieu camera took pride of place and alongside were an assortment of automatic still cameras with high-powered telephoto lenses, a spotlight, and, most curious of all, a parabolic sound reflecting dish. The battery of cameras was centred on a single target so when called to action they would all be filming the same thing. The set up was designed so the operator could shoot with numerous cameras running at the same time; the still cameras were set up with automatic wind capabilities, meaning many frames could be shot in relatively quick succession at the same time as ciné cameras ran. The whole effect would mean catching Nessie with multiple cameras, and, if there were any sound (there had been reports in the past of the monster making a sort of "hissing" noise) then the sound reflector dish would be there to capture that too. It was an impressive piece of ingenuity that in the coming years, like his chin wig, would become part and parcel of Tim's identification amongst monster hunters.

The long engineering sessions which took place in the garage, whether it was working on the Cyclops Rig or, more often than not, the Jaguar, was good news as it gave me a very convenient, and believable, excuse to miss *Doctor Who*.[4] I would argue that Dad needed assistance with whatever task he was undertaking and, just about the time the title music would play, I'd be seen heading for the kitchen and ultimately the garage to offer my services. Now for me it wasn't the Daleks that I found scary; no, it was the Cybermen who had me quaking in my boots. Something about the way they moved and the fact they had no eyes—or just black holes where the eyes should have been—sent my imagination running. Anyway, it was enough to get me out of the warmth of the house and into the chill of the garage where I would pretend to help Dad with all sorts of interesting repairs and things.

4   A popular BBC television science fiction series that ran from 1963 to 1989 and was re-launched in 2005. It depicts a Time Lord, known as the Doctor, who fights a variety of foes while exploring the universe in the TARDIS, a time and space travel machine.

Tim in 1925.

The Dinsdale children, Tim (left),
Peter, Felicity, playing in the garden
in their Antung home, 1930.

The Dinsdale family home
in Antung, China, 1930.

SS. TUNGCHOW

Pirated on January 29, taken to Hongkong on February 1, the Tungchow returned to Shanghai on Thursday morning.

The *Tungchow* about to dock in Hong Kong after being rescued from a pirate highjacking, 1935.

Tim's parents, Felix and Dorys, wait anxiously dockside, their look of relief at their children's safe return quite apparent.

"HERE SHE COMES!"

Tim, in RAF uniform, Rhodesia, 1944.

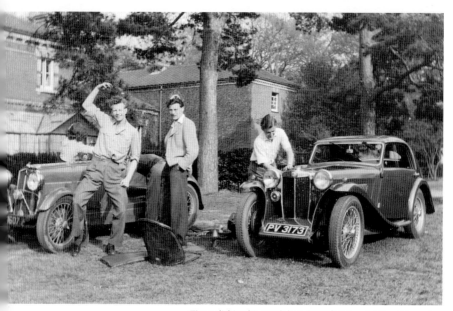

Tim polishing his much loved MG (on the right) amid much
merriment with the boys at Digswell House, circa 1947.

Tim and Wendy marry on June 25, 1951 at St Mary's Church in the appropriately named town of Hitchin in Hertfordshire.

RMS *Franconia* arriving in Montreal in 1951. Wendy was very grateful for the serene waters of the St. Lawrence River after the turbulent seas of the North Atlantic.

Embracing the Canadian lifestyle. Tim's first skiing adventure.

Wendy beating off the chill of a Canadian winter with her new fur coat.

Tim's handy work on display. Built by hand from scratch, the little craft helped ferry all the materials needed to construct the cabin on Lake Kawagama, Ontario in the early days of Tim and Wendy's marriage.

d.

e.

A 14-foot wooden fishing boat tracing the
monster's path one hour after Tim filmed the
animal.

The cone-shaped object moving across the
ch measured 6 feet wide and 5 feet high.

An enlargement.

The fishing boat travelling west, adjacent to
e north shore.

The partly submerged animal put up a 2-foot-
gh wake.

Tim's mobile camera unit the fateful
day he filmed Nessie, April 23, 1960.

To see Tim's film in its entirety visit
www.themanwhofilmednessie.com

Early days for the Loch Ness Phenomena Investigation Bureau.
The second expedition at Urquhart Castle, 1964. Tim on far left.

David James, a founding director and driving force of the LNIB.

Taking advantage of the extra height of the stepladder to view the loch.

Inside the hide on the island.

e crossbow, with an array of flesh-sampling
rts, was part of the Tim's monster hunting
senal from 1966–71. However, a clever
essie kept a low profile during those years!

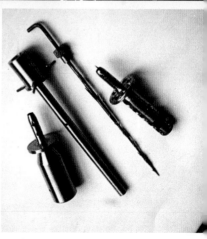

Aerial view of the island.

Tim and fellow independent monster hunter, John Addey, on the island assessing the water after dropping the smelly bait (asafoetida) into the water.

Tim and 14-year-old Simon, looking somewhat perturbed, camping in the ex-army bell tent on the island in 1967, when test firing a flare started a ball of excitement amongst the monster-hunting community.

zara, dressed in her camouflaged
essie-hunting sheeting, moored
ongside the *Narwhal*, at the abbey
er, Fort Augustus, 1968.

Alex, Dawn and Angus feeding swans off
the back of *Cizara* while moored at the
abbey pier, with the catamaran's owner,
Mr. Smith, in close attendance.

Tim on *Cizara* with the
Beaulieu set up on the
Cyclops rig. Tim is sporting
the beginnings of his
expedition chin wig.

The start of the great 1969 expedition, with *Water Horse* in her full monster-hunting regalia with the Moo-scow by her side, stabilizers down and the Cyclops rig with sound dish set to record. Urquhart Castle is in the background. Monster hunting could be beautiful on such a sunny day.

Photo: Ivor Newby.

On the other hand, monster hunting could be miserable. It seemed as though most of the time it was raining.

The deep-sea sub Pisces, commissioned by a film company to tow a model monster, made some interesting discoveries, one being the great depth of the loch, 975 feet. They also had a strange mid-water encounter when the sonar hit a forty-foot "something" that moved off when approached.

American sub mariner Dan Taylor's homemade one-man submarine, *Viperfish*, was easily towed behind a Land Rover.

Dan prepares for the first dive in Urquhart Bay. *Water Horse* is in the background.

Checking instruments at Temple pier, 1969. The man in the wetsuit is Matey, an Australian volunteer with LNI. To the far left is Clem Skelton, the operations director of the LNI before Tim took on the job.

A huge model monster built as a film company prop to use in a Sherlock Holmes movie. Unfortunately, the very day after Tim took this picture it sank in deep water. As one local said, "it went to join its kin folk."

The LNI's workhorse, *Fussy Hen*, tied up beside *Rangitea*. Originally orange, when Tim took over the LNI, the boat w painted in two-foot black and white blocks to provide a si comparison to an object. Should anyone film the monster *Fussy Hen* would be filmed at the same place to provide a accurate idea of the size of the creature.

Aerial shot of Achnahannet, home of the Loch Ness Investigation Bureau 1965–1972, taken by Ken Wallis in his autogyro.

Photo: Ken Wallis

LNI base camp 200 feet above th water's edge.

Simon Dinsdale doing a camera check at the LNI Achnahannet head quarters on an unusually calm day in 1971.

LNI mobile camera unit on site.

Alex Dinsdale, 16 years old, on watch at the Foyers site.

What does a ten-year-old monster hunter look like? The author in a wetsuit after being on harbour building duty.

LNI volunteer Phillipa, Angus (age 11) and Simon checking the camera at a watch station.

The author with an expedition member on watch at Strone, overlooking Urquhart Bay.

Wing Commander Ken Wallis arrives at the LNI
Achnahannet site with the autogyro in tow.

Ken runs through pre-flight checks.

Less than 100 feet of a farmer's field is
all Ken needs to get airborne.

A curious cavalcade heading toward Loch Morar. The big Jag towing *Water Horse*,
an LNI mobile camera unit with the Moo-scow on top, and Ken Wallis's Austin
Mini pulling the eye-catching autogyro.

After a tough day of monster hunting, LNI volunteers would often be seen relaxing at The Lodge Hotel singing along to the weekly ceilidh.

ip Hepple, LNI's Mister Fix It, about depart on a lonely vigil of floating atrol on Moo-scow, staying out on e loch for days and nights hoping sneak a peek at the notoriously y Nessie.

Rip with expedition member, Mac, the parrot.

Brock Badger, a four-season LNI veteran who, after experiencing an underwater close encounter, decided his scuba diving days in the loch were over. Here he is on watch at Strone.

The family at LNI HQ. From left: Tim, Dawn, Wendy, Alex, Simon (back) and Angus (front), with Christine and Brock Badger and a model monster. This picture was taken shortly after Tim learned of Brock's underwater encounter.

Volunteers came from the far reaches of the globe. Australian expedition volunteer "Matey" (right) is helping Tim keep the fleet of mobile camera units serviceable.

1970. Tim, Marty Klein, and Bob Rines (Academy of Applied Science), set off in *Water Horse* to launch Marty's innovation "towfish" side scan sonar. Note the number of cameras in easy reach.

Photo: Academy of Applied Science (AAS)

The towfish enters the loch for the first time.

Opposite, top: Marty Klein, in the cabin of *Water Horse*, studied the side scan sonar results and noticed a large, unexplainable object.

Opposite, bottom: The 1970 side scan sonar chart showing the large mid-water object recorded off the Horseshoe Scr

Photo: AAS/Martin Klein.

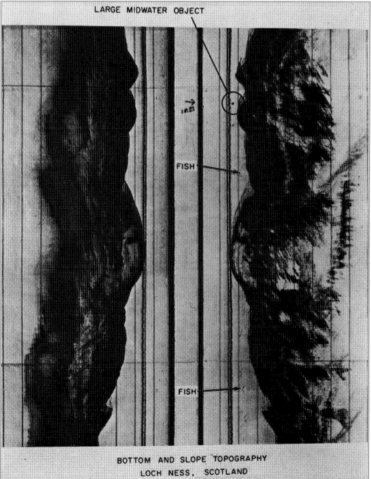

LARGE MIDWATER OBJECT

FISH

FISH

BOTTOM AND SLOPE TOPOGRAPHY
LOCH NESS, SCOTLAND

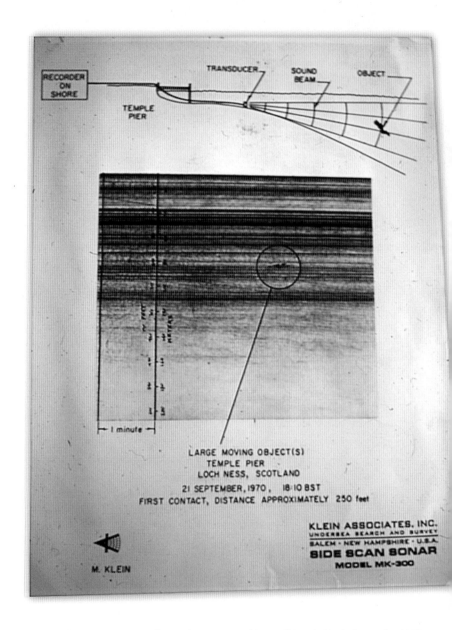

Side scan sonar image showing large moving objects off Temple Pier in September 1970.

Photo: Martin Klein

Academy workers assisted by LNI volunteers carrying the massive underwater sonar flash photography units crossing a field north of Urquhart Castle. Ivor "the diver" Newby is second from the right.

Bob Rines attending his underwater camera.

(L to R) Carol and Bob Rines, Dr. Charles Wyckoff and Bob Needleman lowering the underwater camera rig into place. *Photo: Nick Whitchell*

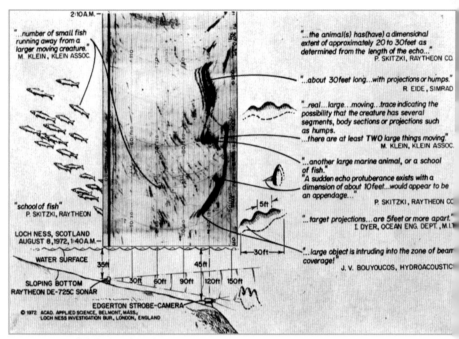

2:10 A.M.—

"...number of small fish running away from a larger moving creature."
M. KLEIN, KLEIN ASSOC.

"...the animal(s) has(have) a dimensional extent of approximately 20 to 30 feet as determined from the length of the echo..."
P. SKITZKI, RAYTHEON CO

"...about 30 feet long...with projections or humps."
R. EIDE, SIMRAD

"...real...large...moving...trace indicating the possibility that the creature has several segments, body sections or projections such as humps.
...there are at least TWO large things moving."
M. KLEIN, KLEIN ASSOC.

"...another large marine animal, or a school of fish."
"A sudden echo protuberance exists with a dimension of about 10 feet...would appear to be an appendage..."
P. SKITZKI, RAYTHEON CO

"school of fish"
P. SKITZKI, RAYTHEON

"...target projections...are 5 feet or more apart."
I. DYER, OCEAN ENG. DEPT., M.I.T

LOCH NESS, SCOTLAND
AUGUST 8, 1972, 1:40 A.M.—

5 ft

30 ft

"...large object is intruding into the zone of beam coverage!"
J. V. BOUYOUCOS, HYDROACOUSTIC

WATER SURFACE   35 ft   45 ft

SLOPING BOTTOM   30 ft   60 ft   90 ft   120 ft   150 ft
RAYTHEON DE-725C SONAR

EDGERTON STROBE-CAMERA
© 1972 ACAD. APPLIED SCIENCE, BELMONT, MASS.,
LOCH NESS INVESTIGATION BUR., LONDON, ENGLAND

The Academy of Applied Science obtained a sonar "hit" in Urquhart Bay in 1972.
The recordings were interpreted as two large (20–30-foot long) "things" moving.
Photo: Academy of Applied Science (AAS)

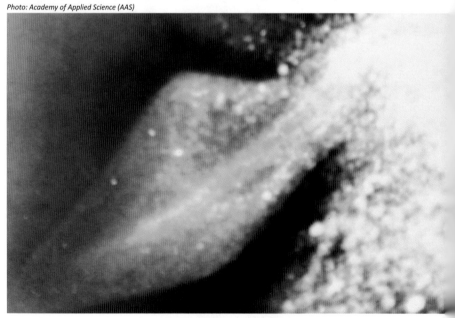

The Academy of Applied Science flipper picture, taken in August 1972 in 45 feet of water in Urqurhart Bay using a strobe camera. The picture shows a paddle estimated 6-8 feet in length and 2-3 feet wide, and coincided with the sonar trace above. The image has been computer enhanced. *Photo: AAS*

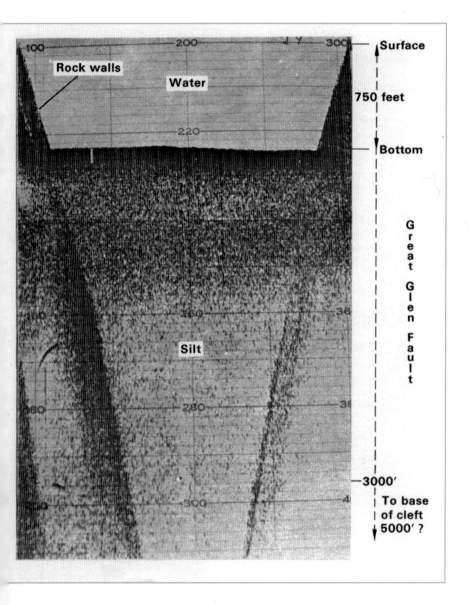

Sonar chart in July 1974 showing continuation of rock walls beyond the deep bottom silt in Loch Ness.

Photo: Academy of Applied Science (AAS)

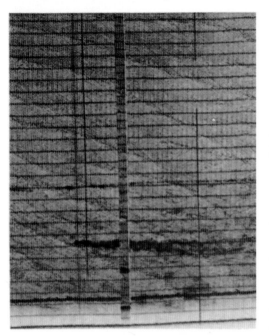

Left: 1978. Marty Klein's sonar unit detected an underwater object. At a distance of 475 feet and a water depth of 705 feet, Gary Kozak and Tom Cummings recorded a target showing a width of 8 to 10 feet and creating a turbulent wake. Subsequent investigation confirmed there was no submarine operating in the loch that day.

Below: The Second World War bomber ditched in Loch Ness while on a training mission in 1940. The plane was first discovered by Marty Klein in 1976 and was subsequently raised and is now on display a the Brooklands Museum in Surrey, England.

Photos: Marty Klein

…he camera crew briefing the author and …ng-time water bailiff Alex Campbell before …e interview for the Japanese documentary.

Alex Campbell being interviewed by the author, then 12 years old.

…m aboard *Hunter*, 1973. Note the …wsprit, ready to go "snout sailing."

Tim with a good friend Father Aloysius Carruth in the grounds of St. Benedict's Abbey. The monks were always strong supporters of his endeavours and gave much practical help by allowing Tim to tie up in the small abbey harbour.

Tim inside Nutter's Nook (his garden office), standing beside the (retired) crossbow. Above is the clay model monster Tim used on the BBC's *Panorama* program in 1960.

Tim, always optimistic of a successful hunt, giving the victory sign while on one of his last expeditions aboard *Water Horse*.

With "Perambulator Three" (P3), Tim returned to his nomadic style of monster hunting.

On the *Eye of the Wind* sailing down the loch. At Tim's feet are his camera case and loose SLR always at the ready!

Tim on watch in later years. The autho took this photo on his last expedition to the loch with his father in 1986

Chapter 9

Aboard *Cizara*

Our second family trip to the loch in mid August 1968 proved to be a great adventure, and provided the rest of the Dinsdale family with an inexplicable close encounter of our own. Setting off in the MK8 Jag, our destination was Fort William on the west coast of Scotland. There we were to meet a pleasant Scotsman named Ian Smith. Mr. Smith had agreed to charter his large seagoing catamaran to the Dinsdale family so we could take her through the Caledonian Ship Canal to Loch Ness. The canal, opened in 1847, was designed to offer a safer route to sailing ships by removing the perils of a journey around the hazardous northern Scottish coastline. The sixty-mile canal joined the east and west coast of Scotland via a variety of locks, lochs, and canals. The most impressive piece of engineering is an eight-lock combination called Neptune's Staircase, which raises vessels seventy feet above sea level at the Fort William end. The canal route passes through Loch Ness, which would be our stopping point, and on to Inverness and ultimately the North Sea. It was a much-anticipated trip and the last real chance for a family holiday, as Simon was about to join the army.

The big car was stuffed with everything a family of six would need for almost a month away from home; plus with all of Dad's monster-hunting kit even the giant Jaguar was feeling the pinch with every available nook and cranny taken up. I got caught trying to smuggle my collection of birds' eggs onboard. I wanted to show them off to my Uncle Peter at whose farm we were stopping en route, however Dad put a stop to that. But for all of his thoroughness a stowaway did make it onboard. Dawn's beloved Panda, originally refused boarding permission, found its way—somehow—snuggled under the back seat, appearing once we were safely underway.

We left home with an overburdened car and the usual arguments about who was sitting where, did we know the way, who was the last person to have the map, and did anyone want to use the toilet once more before we left—last chance!

It was a long and arduous journey, and although the Jag provided a comfortable, smooth ride it still felt like an eternity before finally pulling into the gates of the Northumberland farm. After a good night's sleep we awoke to the joy of a busy farming household. There were the twins, Jan and Bess, who were the oldest of the brood, followed by John, Helen, and Robert. Five of them and four of us; nine Dinsdale children together was quite a recipe for mischief. The two days we spent there were marvellous times. Robert and I teamed up as did Simon and John, Dawn and Helen became firm friends, and Alex with the twins spent hours swapping "groovy" clothes and listening to the Beatles. Uncle Peter showed me what was left of his once impressive birds' egg collection; Robert, his youngest, when aged about three had got hold of them and decided to see if they were hollow by pushing his finger through the lot. The farm cat and dog were characters: the semi-tame ginger tom, Konk, was a legendary rat catcher with one eye missing from an epic battle some years earlier. The beagle, named Benjamin, was a happy dog who knew to only chase the cat when he was sure he wouldn't quite catch him; it was safer for him if he didn't.

We journeyed on up to cross the border between England and Scotland and drove into Glasgow. The Jag, ever temperamental, started to show signs of mechanical problems so we had to endure the heater blasting away on an already scorching hot day. The idea was to take heat away from the engine as it was close to overheating. But it didn't work.

We left the traffic jams and clutter of the big city behind and headed on toward the Highlands and the stunning, dramatic scenery that comes with that ancient landscape. Glencoe is the site of the 1692 massacre of the MacDonalds on a snowy, windswept February night when thirty-eight clansmen were murdered by the very people to whom they had shown hospitality and invited into their homes. A further forty women and children perished due to the bitter cold in the days after their homes had been burnt to the ground. The fields and mountains are steeped in history, and in the failing light with brooding clouds bubbling over the mountain tops one could almost feel those times long past when clansmen would roam the glen, tartan-clad with claymores strapped to their sides.

Having arrived late in Fort William, we had to search for Mr. Smith. We eventually pulled up and got our first glimpse of the much talked about *Cizara*. She was a very attractive vessel and would be our home for the coming days. The inevitable rain started to fall as we packed ourselves in the boat. Exhausted but excited about the next stage of the adventure, we fell quickly to sleep.

We all had our jobs to do and with seven people onboard things had to be kept in order, shipshape, if the trip was to be a success. The first job was for all of us to learn our roles and responsibilities when negotiating the locks. It was a tricky manoeuvre, but once Mr. Smith had explained the ropes—literally—we were on our way. Once on top of Neptune's Staircase we were free to motor off along the canal. As we got more familiar with things, Mr. Smith allowed us to take it in turns to drive the boat. I was so small that in order to look over the top of the cabin and see which direction I was steering I had to jump up and stand on the bench seat that ran along the side of the cockpit. Being the youngest and smallest member of the crew I was under constant surveillance, and my siblings were under strict instructions to keep a keen eye on my every move. But this directive was somehow forgotten one carefree afternoon when mother called all hands to lunch. During the ensuing rush I was overlooked, and as everyone was below tucking into a good helping of bacon and eggs the question was asked, "Who's driving?" Mr. Smith made a pretty quick move from the mess deck to the cockpit to find a happy eight-year-old barely peering over the cabin roof steering his beloved eight-metre catamaran down a narrow canal as if it were the most natural thing in the world. At the time it caused much amusement and I was allowed to carry on at the wheel, but this time with Mr. Smith in close attendance.

As we entered Loch Ness, excitement abounded. Tim had been developing this plan for a couple of years and now here it was all coming to fruition. The catamaran, due to the nature of its design with the twin hull, was a stable boat that would make an excellent filming platform. It made Tim mobile, free to move around the loch at will, accessing areas unattainable from the shore and invisible from the road. Furthermore it was a big boat, which gave a feeling of security and safety.

The loch, at that time, had almost no piers or places to tie up. There were a couple of old wooden jetties dotted around the shore, but these were all privately owned. The problem of where to moor *Cizara* was solved by the kindness of the monks of the abbey at Fort Augustus. Dad had good relationships with the monks and had interviewed Father Gregory regarding a sighting he'd had some years earlier. The abbey grounds backed on to the loch with a small but protected harbour where the water was shallow—one of the very few places on the entire loch—and a stone wall curved out to give protection against the prevailing winds. *Cizara* was tied up next to a derelict longboat called *Narwhal*, a solid boat but in a poor state of repair, which in years to come would have a renaissance and will feature again later in the story.

The weather was being unusually kind to us, sunshine and flat calm water for our first day on the loch. It's what Dad christened "Nessie weather," a term he'd used for some years when the sun shone and the loch was flat and calm, because it felt like the perfect time for monster hunting. We motored away eastward from the abbey pier. After a short while we switched off the engine and tried the sail. Although it was so breathless we hardly moved, it was peaceful and serene. The mountains towering above us showed almost no distinction of a shoreline as the sheer wall of rock plummeted directly into the loch. We came up to a curious formation of scree perched high on the mountainside in the shape of a massive horseshoe, which, as we learned, the area was named. It was quite unnerving straining my neck looking high up as the horseshoe stretched across many hundreds of feet looking like it was just hanging there and could all just fall into the loch at any given moment. There was a bit of commotion as an unusual wave pattern approached *Cizara*. It was like a wake from a small boat yet there was no boat in sight and hadn't been all morning. The really strange part was the small amount of wind was actually blowing in the opposite direction. Not seeing anything resembling a monster, we sailed on.

It was our second day on the loch and the wind was up and so were the sails; catching the breeze, we quickly covered the miles and before we knew it were at Dores at the easternmost end of the loch. Swinging the boat about, we started to make our way into the wind and waves tacking back down toward the shelter of Urquhart Bay. I don't mind admitting my sea legs weren't all that great and I, along with my mother who was never a good sailor, started to feel slightly queasy with the constant swells. Once in the calmer waters we dropped anchor and settled in for the night. Card games and general family fun lead to stories about the monster. As the stories got more descriptive I got more and more unnerved, for here we were sitting in a tiny boat—compared to the size of Nessie according to some of the stories—right in an area where two weeks earlier a group from Birmingham University had some very interesting results tracking something large on sonar moving up and down in about 600 feet of water; and we were sitting right on top of an area she was known to inhabit. Of course I was imagining all sorts of terrible things happening to us and, needless to say, had a very disturbed night's sleep.

Our last day on *Cizara* was taken up with a leisurely cruise back down the loch toward Fort Augustus. We hooked up at the abbey pier later in the afternoon and, after a quick bite to eat, all jumped into the Jag and drove off to meet the LNIB folks at Achnahannet. Dad was introduced to a couple of lads who were staying at the Altsaigh Youth Hostel (which is situated on the shore almost opposite the Horseshoe Scree), who reported seeing a hump moving through the water creating a large bow wake before submerging at about the same time we had experienced the unusual wave patterns the day before.

Later that evening, driving back along the north shore road returning to *Cizara*, we were full of talk about the sighting and what we might have missed. The conversation gradually quietened and we relaxed in to the journey. Being the smallest, I was in the front seat sandwiched between Dad and Mr. Smith. Suddenly, a shout went up from the back. "What's that?!" It was dusk and, as is everyone's wont when driving around this loch, we were all looking out of the windows at the water. At this particular spot the blanket of fir trees broke to give a good, but fleeting, view of the small bay and Cherry Island, a tiny clump of land with two or three trees on it. There were a couple of boats tied up to buoys, but sitting there quite plain to see was the classic "upturned boat" of what we could only assume was the monster's back. Dad hit the brakes hard, bringing the big car to a shuddering halt just behind some trees that obscured the view of the bay. Simon and Dad leapt out, and, grabbing a pair of binoculars, raced back down the road to see what was there. It couldn't have been more than few seconds between the car passengers seeing the "object" and Dad and Simon getting to a position to view the bay, but…nothing. No upturned boat, no Nessie, and, more importantly, no engine noise. It was a still evening and the loch was glass calm. Any boat moving about would have been seen and heard but there was nothing where, just moments earlier, an object large enough to get everyone's attention had been.

We made our way back to the car and drove the remaining five minutes or so to the abbey pier. Alex, Dawn, and I were ahead of the others as we walked down toward the loch. We drew closer to *Cizara* to see her being buffeted by waves, which were also breaking on the shingle beach. This was very strange as it was a calm night. Again, there was no wind or engine noise, so what on earth was making the waves? By the time Dad and the others arrived, *Cizara* was still moving about at her mooring, but nothing like a few moments earlier. Dad and Mr. Smith were perplexed at what could have caused the big catamaran to pitch up and down so much. As the light of the day had now all but gone, Simon made the suggestion to let off a flare, which would enable us to see what was out there in the bay. The train of thought was that perhaps the monster had swum out past Cherry Island and into the deep water coming around toward Borlum Bay[1] where *Cizara* was moored. If we let off a flare we'd get an illuminated view of the entire bay area and perhaps a sighting of Nessie. But Dad was very reluctant to do so; after all they are universally recognised as a signal of someone in distress and with the township of Fort Augustus right there, it could easily set off a false alarm. I also think that perhaps because of the happenings of the year before when, on the island with Simon, they "test fired" a flare that set a wheel of false understanding in motion, Dad didn't want to repeat the episode. We contented ourselves that we had had a very strange but real experience; everyone in the car except Dad, who was keeping an eye on the road, had seen something. Mr. Smith thought it might have been a boat but admitted to being at a loss as to explain how it had just disappeared. We spent our last night aboard excited and thrilled to have our own inexplicable Nessie story.

The following day Wendy and the children left to catch the train south. The kids had to get back to school after the long summer break and Simon had a date with Royal Armoured Corps. Tim stayed, taking sole responsibility for *Cizara,* as Mr. Smith also left. One can only imagine his feelings as he motored off leaving his prized yacht in the hands of a novice sailor.

1   Borlum Bay is at the westernmost (Fort Augustus) end of the loch. A crescent shingle beach slopes into the loch, descending to depth quickly. The River Taff enters the loch there, as does both the Caledonian Ship Canal and the River Oich. The Benedictine Abbey, and small harbour where Cizara was moored, is sandwiched between the two rivers.

Chapter 10

Alone On the Loch

Tim set about making the boat ready for single-handed Nessie hunting. Now that he had more space, the Cyclops Rig was set up in its full glory with the sound dish and all the odd gadgets specially designed for the sole purpose of trapping Nessie on film. Tim had long harboured the thought that the glaring white of the boat's fibreglass hull could be something to scare Nessie if she was to surface nearby. Certainly, she was reported to react to sound, so why not sight as well? In the months leading up to the expedition, Tim had devised a plan to combat this very problem. He had made a camouflaged "skirt" to drape around the hull of *Cizara*: he measured and cut up some old bed sheets and then, using Wendy's tub washing machine, started to dye it all khaki greens and browns. This act was almost the last Tim ever undertook, as it was just about a step too far for Wendy. To say she wasn't impressed would be a massive understatement. She had put up with much over the years; a huge drop in income, looking after four children while Tim was away on expedition, and suffering the occasional ridicule surrounding the subject were things she understood and accepted as part and parcel of supporting Tim in his quest, but this was pushing it just about as far as it could go. Needless to say, the washing machine was never again used for such a task; and although *Cizara* did get her new gown and wore it proudly during the weeks that followed, the line between monster and family had been firmly drawn.

The hunt continued for a couple of weeks with Tim alone aboard the big yacht. His nervousness at handling such a large craft single-handedly comes through when reading the letter he sent home in the days that followed:

Ft Augustus, Saturday, Sept 14th 1968.

I thought you would like to know how things are progressing…

When you left on Saturday morning last, I must say after all the noise and activity and squash aboard I felt pretty lonely, and the slow process of getting things into some sort of order was started without much enthusiasm. I decided to stay at the quayside—or rather tied up at the Abbey Haven, which is what I was now calling it. For some reason I was feeling tired too, and kept knocking off for a stretch out on the bunk. On Sunday, these stretch outs became a necessity, because I began to feel most peculiar. A thumping headache, and a curious lassitude, just as though I'd been dipped in treacle; it was a huge effort to do anything at all. On previous expeditions I have noticed this—perhaps a sort of physical boomerang to all the weeks of preparation and worry about equipment and forgetting things, etc., only this time it was worse.

However, by Monday I'd got over it and the long job of dressing *Cizara* in her camouflage skirt was done, and very satisfactorily too. These long sections of old sheeting dyed and painted in blobs are tailored to the hull on either side exactly, and suspended by lengths of sisal string. The brilliant white of the hull would certainly catch Nessie's eye if she has any sort of vision at all.

Another job, which turned out well was the simple idea for a dark room. I draped the old blue bed cover over the windows in the toilet, and although of course if doesn't keep out all the light, its subdued enough now for me to change films.

Regarding the very unwelcome "boarder" we had on Friday night [someone had tried to climb aboard presumably to see what was easy pickings to steal], I mentioned this to Fr. Aloysius…I think the tuneless singing we heard was a ruse to see if anyone was about. When there was no response, the man decided to come on deck to find if there was anything worth lifting. I told Fr. Aloysius I would not allow anyone to come aboard Ian's yacht without permission, and meant it—and he quite agreed—but in the interest of preserving the peace should a second attempt be made, at night I string a long line and pull *Cizara* out into the water. Anyone who wants to board her in the future will have to swim first before receiving a large biff on the nose—but there's been no trouble since.

On Monday afternoon, I decided to make a sortie, but when I went to start the engine—grr grr and then nothing. The battery was flat. I had to start it by hand, which was quite hard work, as it has a very high compression. I got away all right and cruised around for a bit, experiencing that mixture of elation and anxiety which is a part of every first solo performance—be it in an aeroplane, on a push bike or a pair of roller skates…A small naval vessel went by towards the Caledonian, and paid attention through binoculars. I felt very "lone sailorish" and toyed with the idea of dipping the Ensign—only I'd forgotten to put it up.

It was quite calm, and after buzzing about to bump up the battery I nosed gingerly into Cherry Island Bay, where you all got your monster sighting on Friday night, sounding out the bottom on the fathometer. There were a few boats and yachts moored there, next to a pier, quite obviously in use (which is the exception on LN) so I guessed there must be a good anchorage. The machine showed ten feet under the hull. I should think it is the best anchorage on the Ness.

I got back with some difficulty as there was a side wind, and I over corrected and missed the first approach. As you know, this is a safe little harbour, but there is no room to spare sideways between the main protecting jetty and the partly completed one, sticking out towards it from the opposite shore. Anyway, I made it and felt much the better.

On Monday night there was a brilliant steely moonlight, and I awoke to the screech of a Loch Ness squall in the rigging. The shore ropes were bowstring tight, and dear old broken-down *Narwhal* with her crazy bowsprit askew, cavorted back and forth clinking her piles of rusted chain. *Cizara*, not to be outdone, joined in the dance, a kind of minuet it seemed to me, bobbing back and forth sometimes in harmony, and sometimes out of step. To and fro they went, obviously, enjoying it. I watched until two in the morning then dozed off to a dull and cloudy morning with a tinge of pink in the sky.

I got on with the chores, then I tried the engine again. It barely turned over on the starter, then refused to do any more…so I started it by hand.

I set off with misgivings. The wind was rising, and I knew if I stopped the engine I would probably have to start it by hand, and if it didn't go I might drift on to the lonely rock-lined shore before I could get the sails up: allowing twenty minutes to do this for a single-hander.

About two miles out I stopped, and began to drift about. The water was black like ink with the heavy overcast, but with patches of intense ripple on it, like vibrations ripples almost: it looked sinister and I found the boat drifting along at two or three knots without any sail at all. Another curious thing—for some obscure reason, due perhaps to the unusual light the vast sheet of water seemed slightly convex, like the meniscus (or is it miniscus) on a tube of mercury. I decided to try and start the motor, just to see if it would go…and when it did, on the button, I was so relieved I found the boat heading straight for home and safety with me holding the wheel. But it wouldn't do. I had a lot of things to test, so I stopped the damn thing again.

On Wednesday I left the Haven before breakfast, and went out to the Horse Shoe. It was a mist-enshrouded morning, with the water flat calm. I let go of the helm and drifted about becalmed for about six hours without touching it. Before me the loch stretched to the horizon, with the mountains draped in swaths of mist. Not a boat or a human being in sight—just the gigantic sheet of jelly with weird reflections on it, and *Cizara* alone on the surface.

On Thursday, by contrast, I set off and motored up to Foyers in blustery wind, with whitecaps everywhere. It was just like being out at sea. I got there at 2pm, scouted the bay, then went out to the exact place on the water where I saw the Beast in 1960. I wanted to see what a human being would look like on the little road high up on the south shore, at the place where I had been that eventful morning. A sort of 'Nessie's eye view,' so to speak.

Shutting down the motor, I turned to drift home, a distance of about 12 miles. Cizara made three knots on the log, heaving up and down gently like a great rocking horse. There were whitecaps as far as I could see, and the magnificent scenery bright in the sunlight. What a wonderful experience!

At first the boat yawed from side to side, but surprisingly, I found I could control her on the rudders and tack her as though she had sail up…only without the mad scramble of going about. Sailing without sails, in fact!

And so all the way back to the wee harbour in about four hours; but here I ran into trouble. Trying to squeeze in I was broadside the wind. *Cizara* drifted sideways to within inches of the rocks. It gave me a real fright believe me. But I made it on the next attempt and leapt ashore to do battle with the ropes. Don't ask me how I won the tug of war, because the wind force was terrific and the catamaran nearly pulled me into the harbour.

Curiously enough on the way in that day I had a presentiment of danger—of something going wrong, and the need not to venture out on Friday. Quite obviously unless the wind died down I couldn't, but it was also to be Friday the 13th which as any sailor knows is a good enough excuse…

Cizara was returned to Mr. Smith undamaged and in perfect working order; but while Tim was relieved to be relinquished of the responsibility, he was smitten with the waterborne expedition. The vast loch, stunning in its beauty yet scary in its moods, had captured his imagination and on the long drive home to Reading his mind was busy working on plans for the following year's hunt.

Chapter II
Subs & Sonars

The winter, as always, was the time for plotting, planning and scheming. Saddened at the lack of tangible results in 1968, Tim kept his attention on the intangibles—ongoing sighting reports from credible folks, not least the Dinsdales' own experience and the exciting sonar hit from the Birmingham University crew—to sustain his belief of success in the coming season.

Like some of Tim's equipment, the expedition ideas became more elaborate. Operation Albatross was a plan to use a powered glider to patrol the skies over Loch Ness. An observer would keep an eye on the waters and if anything needed to be investigated the glider pilot would switch off the engine and swoop silently down to capture its prey on film. In principle it was a good idea but not all that practical and so, for the time being, shelved.

Tim's reputation as a lecturer was growing. He did no advertising, relying instead on word-of-mouth to gain his engagements. The Universities of Liverpool, Wales and Queen Mary College London had all sent invitations, and this was a clear indication that the subject was reaching a higher level of acceptability even if the official scientific bodies continued to ignore it. The Birmingham University sonar results had caused a stir and, yet again, the question was being asked, "What the devil is it?"

Tim was busy trying to secure another boat, either as a charter or purchase, in preparation for the summer season of 1969. There was some talk of the LNIB helping out with sponsorship, but in the end it didn't materialise. So Tim went searching for something to suit both his expedition needs and his pocket. After viewing a number of likely candidates, a smart sixteen-foot, one-person cabin cruiser was decided upon. It was tiny in comparison to the luxurious *Cizara,* but Tim felt it would suffice for the role it would be required to play.

Water Horse, as she would soon be christened, added to the already bulging array of expedition equipment. To make room for the boat's arrival, Wendy had to sacrifice yet more of her ever-shrinking front garden. Part of the fence was removed and the front lawn shaved at an oblique angle to increase parking space. The Jaguar, Mini, and now a boat filled the driveway, making quite a sight for passersby and provoking a few comments from the neighbours.

She was a slick and happy-looking boat. Small and with smooth lines, she was obviously built for speed. There was a compact cabin with just enough space for a cooker and a couple of bench seats, which had storage space beneath and converted into a comfortable bunk for one. While nothing about her was spacious, the cockpit area would be sufficient to house the Cyclops Rig and still allow Tim to move around.

Shortly after her arrival she was quickly pounced upon by us kids; we had hours of fun clambering on, in and over her. However it was to be short-lived, as she became the next addition to the "very important piece of equipment" list and with it our nautical playing days were greatly curtailed. Not that Tim was being a killjoy; the boat was sitting on a trailer and numerous little feet clambering all over were actually doing some damage to both boat and trailer.

The small boat had a narrow hull, meaning it would have none of the stability characteristics of the big catamaran; this would be a problem because as the swells at the loch buffeted her to and fro, the movement would render her next to useless as an effective filming platform. It was time to go back into the garage and work on yet another ingenious design. This one was to be quite outstanding, even by Tim's clever standards.

Using the rusting remains of the children's old garden swing, Tim built two outriggers. Attached to either side of *Water Horse,* the arms stretched out horizontally about five feet, and vertical sections went down below the water line where each arm had a large dustbin lid attached. It sounds really quite odd but was, in fact, rather effective. As the boat pitched and rolled the dustbin lids acted like hydraulic dampers lessening the undulating action; it wasn't perfect by any means but it would certainly help with stability. Once the whole set-up was painted black and attached, it all looked nothing more sinister than a couple of fishing rods poking out each side of the boat.

Besides the odd equipment and slightly bizarre inventions, the business at the loch was serious. Tim was not only a novice sailor but also a real beginner when it came to boating in general, and he knew he was pushing his limits when it came to living and working aboard such a small craft. Loch Ness is a huge body of water, twenty-four miles long, a mile wide, and with such great depths that the area generates its own weather patterns. Wind whistling down the Great Glen condenses, creating a vortex that whips up a storm turning the loch's smooth surface waters into a boiling mass of white-capped waves in a matter of moments. With little or no shelter a small boat can get thrown dangerously around. Trouble, real trouble, can spring from nowhere in seconds. Tim had a taste of the loch's moods and the speed of how fast things could go out of control while he was single-handedly sailing *Cizara*. The catamaran was in a different class compared to *Water Horse;* the bigger boat would ride the waves comfortably, giving a general feeling of security, while something as small as *Water Horse* would be tossed around, bobbing like a cork on an angry sea.

It was to be a steep learning curve and, as there was no manual or "Idiots Guide to Boating on Loch Ness" for reference, this was a learn-as-you-go situation, one that would leave little room for error. If Tim got it wrong it could cost him his equipment and perhaps even his life.

Preparations for the 1969 expedition, his eighteenth, continued throughout the spring. A delay in departure, due to having the wartime bullet removed from his right hand, meant missing the first month of the season. The Jaguar had spent time in the garage with Tim customising it to tow *Water Horse*. Once everything had been packed and was all ready to go, the whole set-up looked very smart indeed. Tim had gone from a borrowed camera sitting in his car where the front seat should have been to a fully self-contained Nessie-hunting unit: car, boat, crossbow, cameras, the lot. He was ready once more to "do battle" as he would say, with both the loch and its monster.

The family had accompanied Tim to the loch the previous two summers so was happy to let him go off alone on this hunt, the longest by far to date. Plus, Wendy had an appointment with royalty. She had been asked to attend a garden party held in the grounds of Buckingham Palace in the presence of Her Majesty the Queen. The invitation had absolutely nothing to do with her husband's unconventional occupation; Wendy was to represent the Citizens Advice Bureau, her employer, at one of the Queen's annual summer functions. However the July date was slap-bang in the middle of Tim's Nessie hunting season so they agreed to go and each have their own unique experiences, Wendy's in the heart of London, Tim's in the Highlands of Scotland.

This was to be a season like no other because the national media had widely reported the Birmingham University sonar results creating a lot of interest from a whole slew of groups, institutions, film companies, and a growing list of individuals intrigued by the mystery. The drama of it all, and Tim's involvement, is best described in a series of three letters he sent home to the family throughout those summer months:

Aboard *Water Horse*, Urquhart Bay, July 15th 1969

Dear Everybody,

This is the first of the letters I will write to keep you up to date with what's going on.

After leaving you all early that Sunday morning, I motored northwards, feeling a little sad, but in compensation the trip through the country lanes of Northamptonshire was lovely. The boat trailed easily, though it did not take to rough roads, giving one the impression it was connected to the car with a thick strand of elastic. I think it may have been the bumper bending, but as I could not actually watch it, I ignored it, and periodic checks showed everything to be in order.

I arrived at Peter's in the mid-afternoon, having covered 330 miles, which says a lot for an early start. They were all in fine trim at the farm. I spent a welcome day with them, then left on July 1st, crossing the great suspension bridge over the Firth of Forth, and so northwards to Inverness.

I drove to the Caledonian Ship Canal at Clachnaharry to make my number with the manager there, and from him I learned of the excitement going on at the Ness, with the two submarines at Urquhart Bay, one belonging to Vickers, a sophisticated research vehicle called Pisces; and Dan Taylor's little one-man job Viperfish which had yet to be launched.

On to Fort Augustus, in brilliant sunlight, I passed the spot near Cherry Island (where you got your sighting last year) and examined the launching ramp. It was in need of repair, but I reckoned I could just manage to slip *Water Horse* there. Had it not been suitable it would have meant going back forty miles to the Muirtown three-ton crane, and lifting the boat into the canal, and so into the Ness.

Having done some shopping in F.A., I went back and started preparations, only to find the job impossible on my own. It proved far more difficult than I expected; to jack-knife the car and boat into position, and with so much at stake I couldn't risk an early disaster. I phoned the expedition [LNIB] and asked for a crew to help, but as it was such a fine day most were off on camera duty, so I was told it might take an hour or two to round them up. I set to work stripping the car and boat of gear, ready for the launch. Perhaps an hour later the proprietor of the hotel, nearby, came down and offered to help. It was a tricky operation, but we got *Water Horse* into the loch safely.

I spent the night tied up alongside, surrounded by the chaos of equipment, and the next day, which was wet, tried to sort things out. I just could not see how to move in such a microscopic space—compared with *Cizara*, which is a floating hotel by comparison. Late that evening, I set off for the tie-up point up the river Tarff, promised me by the monks, but it was dismally cold and wet, and approaching the river mouth *Water Horse* went hard aground on what appeared to be a ridge of stones blocking the channel. That gave me a jolt, and for a ghastly moment I stood waiting to see the loch pouring in. I managed to pole off with an oar and went back to the little bay, and spent the next four days and nights there, anchored, away from the pier with its clatter of youngsters and fishermen leaping in and out of boats.

It was a lonely time, and a damp one, and I struggled to keep some sort of order aboard. Obviously the expedition was going to tax my patience. Every move I made had to be a conscious one if I was not to bang my head or sit on a camera or some other vital piece of equipment. But it wasn't all bad; the cooker was a success, my bunk was okay and the visibility marvellous from inside the boat. Once I got the gear into corners and nooks, stuffed up the bow and under the seats—and remembered where the hell I had put it—I could begin to move about. I decided, too, to go through the boat with an eagle eye, and transfer all non-vital equipment back to the Jaguar.

By Sunday, I was beginning to see daylight, and had found that the 'dustbin damping' outriggers were a stunning success. They damped the small-boat wobble to a degree that made a camera platform comparable with *Cizara*.

I left Inchnacardoch on Monday, a week after getting to the Ness, and set course for Urquhart Bay, having parked the car at the abbey—confidently expecting to return within a day or two, once the sub was launched. About a mile from the starting point, with a strong wind, the engine faded out and *Water Horse* began to buck uncontrollably. I realized that small boats don't behave like big catamarans and hurriedly tried to start the engine again, fearing the worst.

It started at once, to my surprise, and buzzed away contentedly for the remainder of the journey, at full throttle, pushing along at about 8 knots. I think it must have been a spot of water in the petrol condensing off the tank.

Opposite Invermoriston the wind increased, and in no time we were in rough water. I kicked the driver's seat down, and stood at the wheel with legs braced. We surged forward and as the wind increased I noticed that *Water Horse* was putting her nose down and surfing! It was an exhilarating ride, and so lonely, with nothing but the rock-walls towering above on either side. The sun glinted off the wake and the pursuing waves, and I made a foolish attempt to catch it on film, letting go of the wheel. The boat immediately broached, and for a moment I thought she would roll over. It gave me a fright. Four or five miles further in, the wind abated and the rain began to deluge in icy sheets which swept across the water. I put on my "oilies"[1] and stood with my back to it comfortably enough, and in time it stopped, leaving only an inky surface with the swell coursing through it.

Perhaps an hour later we rounded Castle Urquhart, entering the bay in sunlight. It was sheltered and calm, and I tied up at the small floating LNIB jetty, close to Temple Pier. There was a great deal of activity, with a film company occupying the shore, building a gigantic model monster for use in a new Sherlock Holmes film; next to it the bright orange submarine Pisces, with its attendant score of technicians; I was told they expected to use it to propel the model monster, once they got it in the water.

---

1 A slang term referring to fishermen's waterproof clothing, oilskins.

I was also told that, due to a technical snag, Dan Taylor's little one-man sub would not be launched until the next day, so the throng of sightseers would be disappointed, not to mention the press photographers. I went ashore to visit Urquhart Lodge, or rather the garage behind it in which the submarine was parked. It was an odd sight. A tiny yellow cigar, made out of fibreglass—hundreds of layers of it, with all sorts of external cables and equipment and a conning tower with thick observation portholes. It may have been a do-it-yourself job, but one got the impression the owner knew what he was about. I met Dan briefly, and liked him—a quiet southerner. Equipment was everywhere…

I spent the night tied up at the pier comfortable enough, except for the thumping noise generated by wavelets as they collided with the hard chine of the hull.

The next day brought the press back in droves, with frogmen and local boatmen and hundreds of onlookers and cars causing a traffic jam above. Fortunately the sun came out, and a further launching delay of 24 hours did not upset anyone.

The day after that caused a disturbance which must have been unique to this part of the Highlands. Dan Taylor's sub was towed down from Drumnadrochit on its curious three-wheeled trailer, and hoisted up by the crane used for lifting Pisces in and out of the water. The shore at Temple Pier, which is privately owned, was seething with pressmen, the film company employees, the LNIB volunteers, and the tourist multitudes above. On the water frogmen darted about blowing spouts of water out of their schnorkels, and on the surface boats jockeyed for position. It was a form of three-ringed circus, but when the little submarine refused to submerge because of ballast problems it was decided to postpone the operation.

For the next week I moved about the bay, driven from one spot to another by the ever-changing wind and the small boat's need for protection. I seemed to be constantly up-anchoring, and with the light remaining until midnight, and dawn at 4am due to the northerly latitude, sleep became a dreamed of luxury.

Aboard *Water Horse*, Urquhart Bay, August 12th 1969

Perhaps I should say half-time letter, because I've been on the water six weeks already, and the days and weeks slip by in such an extraordinary way I often forget what day it is. I was once three days out, and I'm sure the local shopkeeper who put me right thought I was barmy: I said, "Is it Wednesday or Thursday?" when it was Saturday afternoon!

This may give you some idea of the dreamy surroundings in which I have now come to live, because after the early mistakes I settled down to a quiet routine which might best be described by the phrase 'Where'er the wind bloweth—that's where I goeth.'

At the moment I'm anchored up the delta end of the bay, and the wind bloweth—or is beginning to, and, as typing with the boat going up and down is difficult, I'll have to move, using the electric outboard which moves me with stabilisers down in silence, at about one mile an hour...Ah that's better. We are now almost in among the reed beds. It is a lonely and lovely part of the bay with, curiously enough, a shelving approach of sea sand infiltrated with river silt. Shades of the past, and not so long distant either, when the Ness was an arm of the sea. As you may remember from literature, the Beast has been reported out of the water here too, though I know of no recent accounts. It could certainly get out here easily enough, and I watch the shoreline for signs of it.

To go back to where I left off—very briefly, I came to terms with my situation about the second week in July, when I visited the Carys high up on the west shore of the bay. They have an old converted croft house, and several acres of steep hillside leading down to the water in the vicinity of Urquhart Castle. They are retired folk of 'boundless hospitality' to whom I referred in *The Leviathans*[2]; and they haven't changed a bit. Basil soon suggested putting down a mooring for *Water Horse* close under the lee of the shore at a spot called Goat Rock, and with the prevailing westerlies I hastily agreed, because it meant I would get some proper sleep. It was away from the constant activity and chatter of the pier across the bay.

Within a day or two a routine developed, which I have followed ever since. I get up with the light, photographic light that is, which is about six, set up the gear on the big tripod, which is strapped down to the hull, and watch in absolute quietness until about nine. I tidy

2 Tim's second book, published in 1966, which examines reports of lake and sea monsters from around the world.

ship, which is difficult in such a confined space, then slip moorings, and allow the wind ripple to take me hither or yon. Sometimes I drift out into deep water in the loch proper, and sometimes just potter about in the bay. And then an hour later on most mornings a wind gets up—I call it the 'ten o'clock wind.' With stability lost I motor over to the LNIB pier and exchange pleasantries with the overnight camera crew, who have kept an eye on Dan's little sub, which sits on its three-wheeled cradle just clear of the water. The cradle is winched up and down a shelf of stones and the submarine floated off for diving operations.

Of course the ten o'clock wind isn't always on time, and when this happens I continue the drift until it does appear, except in 'Nessie Weather'—the hot flat calm we all hope for, when I drift about for most of the day.

In the early part of the hunt, Pisces was diving a great deal, and making almost daily discoveries. Huge potholes and fissures in the bottom of the loch, confounding the echo search equipment from the surface, which had plotted a flat, almost featureless plain of silt. They shot film of it, and discovered too an old wreck of a ship off Temple Pier, some guns or muskets which they could not quite pick up with their grab, and two places of great depth, beyond the previously accepted maximum, which had stood at 754 feet off the castle, since the loch was first mapped underwater. The echo-sounder new depth of 820 feet and then an abyssal 975 feet near the first Cobb[3] marker post: the place where *Crusader* started her ill fated record-breaking run 1952. They had two other experiences of note, as you must have heard by now: the big blip on the sonar screen, about 40 feet in length, by interpretation, which shoved off when they approached it, and a strange vortex which spun them round in a big hole far down near the bottom. The submarine weighed eleven tons, but was swirled around, and had to blow tanks to get out of it. I think it frightened them, because it could have been caused by a current, or vortex, or possibly a sinkhole in the bottom of the loch with water running out of it. If the latter, they might have ended up sucked into it like a bath plug—for keeps. Being a fault zone, the whole length of

3 John Cobb was a famed racing driver and speed enthusiast of the 1930s, 40s and 50s, who died in 1952 while attempting to break the world water speed record in a jet-propelled speed boat, *Crusader,* on Loch Ness.

the Great Glen could be seamed with cracks and fissures underwater, as it is above.

Another diversion was caused by the film company and their monolithic model monster, for use in the Sherlock Holmes picture (though how it has found a way into the Holmes saga defeats me). They spent a lot of time and money towing it about, and making artificial mist to film it through. The bay was alive with boats for a while, and the constant disturbance of engines, which of course put an end to monster hunting—the real kind that is. But it was fascinating to watch.

The model was about 30 feet in length and weighed four or five tons. It had three huge humps and a long neck with a ferocious head, and bulging eyes. It could move its head about and open its mouth, and the eyes lit up too. I believe the whole contraption cost about £10,000. I was lucky to film it from the western hillside one day when they started tests. A cable was secured to it from the shore, then a power winch dragged the 'Beastie' through the water at quite a speed, which with the vane beneath, produced a diving effect, the head and neck moving all the while.

I saw this, and having the Bolex camera with me flung myself down and, resting the lens on the carrying case, shot off some film. I then slithered down to the shore, and shot some more film in close-up as the model went by. A boat appeared with frogmen, who got on to the model, then in through a small trap door in its back, rounding off the sequence.

The very next day the great monster plunged to the bottom of the loch in rough water, and was lost—or as a local newspaper put it 'went to join its kin folk.'

Continued 15th August which is, I feel sure, a Friday—or is it? But no matter, I am out in the deep water in perfect calm, with mist veiling the far mountains of the south shore. It is fantastically beautiful, with a sort of cascade of sunlight streaming down through it and gleaming off the water. The castle is to starboard. Wee *Water Horse* now looks odd, draped in camouflaged sheeting, cut up from the shrouds around *Cizara,* but it is effective, and blends with the shoreline. Indeed, the other day nature paid us a compliment. A flight of wild duck nearly flew into us—and for weeks now I have been watching and filming a family of Great Crested Grebes, diving birds, nearby. They were chicks when I got here, but now are twice the size, and they cover great distances each day, thinking nothing of swimming across the bay and back, a distance of a mile and a half at least. They are pretty things with pointed beaks, and a ruff of feathers to the side of the head. They can also run on water. Honestly, exactly that, and very quick too, for perhaps a hundred yards at a stretch. They are still too young to fly, and I have tried to film them in slow motion doing this V-wake like a small flotilla of speedboats. They absolutely pelt along, then dive as a final resort, I don't like to frighten them, so veer off before they run out of breath…

But back to the model monster, and the many other things of note. The film shot came out well, in fact it must be unique, and it may be invaluable in providing a direct contrast between the most expensive model Nessie ever built, and, perhaps, the real thing; but we must wait and see.

Altogether I've shot about 1,000 feet of film as backing—and nearly all of it appears to be good, with colours well in balance; which is an encouragement. It has given me some priceless scenic effects.

But it is not all dreamy. In fact I shall remember the expedition for the great variety of human contacts. The LNIB fields a team of about 20–25 people each fortnight, and when I go to the pier I meet them, or rather two or three who guard the camera there, and the sub. It is nice to get ashore periodically and talk to folk, and inevitably there is much fun and good humour. I get water there too, and am able to arrange a swap of batteries for the electric outboard. Dan Taylor is, of course, constantly charging batteries for the sub.

Viperfish, by the way, is now operating and it dives to the bottom in the bay. Dan was down the other afternoon, and had a big trout swimming round the portholes in his conning tower looking in. He has lights which help a bit, but the water is so ink-like even Pisces with her terrific lights can only see a few yards at best. Underwater, sonar is the tool to use, and we waited upon the arrival of Bob Love, the American underwater search expert with his box of tricks; and the Birmingham University team, and Plessey's waterborne equipment to be mounted in *Jessie Ellen*. All this seems to be scheduled for September. Meanwhile Dan is working up his boat, and doing everything sensibly and quietly. He is levelheaded, and in my view has a lot of courage. The press seem to have got bored, and do not bother him nearly as much. In fact, the pier end of the bay is now relatively quiet, with Pisces away tomorrow to start diving in Loch Linnhe, which is salt water. She and her crew will be missed, I'm sure, particularly Bob Eastaugh her skipper, who is a big strapping fellow. The sub-mariners are a breed of men apart, and I admire them, but for some reason the underwater search does not draw me like photography.

The ten o'clock wind is here (about an hour late) and I'm drifting down towards the Cobb mile post again. It is a lonely spot—perhaps harbouring the shadows of disaster. I think I'll move away… Ivor and Christine in their big glass powerboat, with Ron, another monster-hunting independent aboard, have just tied up alongside for coffee. I'm not bothering with surnames, because no one seems to use them and it is quite impossible to remember them all—anyway, they moved into the bay about a week ago, and are doing much the same as I, but without all the falderal of equipment. It's nice to have some waterborne company, and we often tie up alongside and drift. His tiny dingy, a sort of li-lo affair which we call the 'Banana Boat' is even more easy to fall out of than the Moo-scow, which is the name I have given the inflatable dinghy which trails around after me. It reminds me of a large and rather dim but friendly animal which wants desperately to please, but sometimes does the opposite. Sometimes, I'd swear the thing was alive. It casts itself off and drifts away, and when I go after it, the moment I try to and catch it with my boat hook, the breeze freshens and it eludes me. It's quite a jolly old thing really and is indispensable for getting ashore, and as a safety boat—or life raft.

Aboard *Water Horse*, Thursday Sept 11th 1969

I'm sitting in 'me cabin,' looking out across the bay through the glass windows—having just cooked a very pleasing lunch of lamb chops and peas, swilled down with Pepsi-Cola!

There is a strange collection of boats around the LNIB floating pier, set off by the backdrop of the far shore, the castle, the rough water, the towering south shore mountains and the grey leaden sky. It is cold, but the rain has stopped, thank goodness. My worst enemy in the very little boat.

Perhaps a hundred yards away there is an old rowing boat: black clinker-built hull, with a green paint strip round the top. A man is standing up, fly-fishing. He is old too, with white hair, and his body marvellously upright compensates for the pitch and roll of his craft. The fly rod bends in a gentle arc to the movement of his hand.

Astern, Ivor Newby's gleaming white cruiser trails the Banana Boat, which has a nosey look about it—a longish snout which twitches up and down with the movement of the water, and two rubber-patch eyes set too close together. For some time now we've noticed the Banana Boat making up to Moo-scow, and when hitched up together he jostles her. The Moo–scow is more than twice his size, and pays no attention whatsoever—or could it be her mental faculties aren't up to it, that she doesn't even notice the intrusion? Experience tells me this is probably the case…

On Temple Pier nearby, where the Great Model Monster sat, the sonar teams from Birmingham University, and Plessey and the Navy are preparing their equipment, with ITN[4] and the *Daily Mail* in attendance. I don't know what results they'll get, but one thing is certain, the hunt is almost up for me, and on Sunday week I will move out of this bay which has provided shelter, and so much else besides during the past three months. But it is not over yet, and in order to provide some continuity I must tell you a little of what has taken place since I wrote the half-time letter.

During the latter part of August the hunt was galvanized by the arrival of the Petersons: Ken, the Producer for Walt Disney Productions, with whom I had exchanged correspondence, and Harriet, his wife. He came here with the object of shooting colour for a short TV film, and I'm sure he and his expert Scottish camera team have made a brilliant job of it.

The LNIB crews and just about everyone concerned this summer became involved in the fascinating game of film making, with the loch providing the best variety of moods and colour as a backdrop.

I think the Petersons enjoyed their fortnight very much, and went home with some unique film, covering all aspects of the chase: the witnesses, scenery, songs and entertainment at the 'Lodge,' the meeting ground for monster hunters in Glen Urquhart.

The Lodge, by the way, provides a spontaneous form of entertainment once or twice a week, known as a "kaly" (spelled 'ceilidh' in Gaelic), at which kilts are worn, the pipes are played, and where individuals sing and dance into the early hours, prompted by a dram or two of that Scottish stimulus known to some as the 'water of life.'

Looking back to my jotted notes for other points of interest, I see that on August 17th Ivor Newby 'falls overboard for the 4th time; clutching anchor.' Falling in is more easy to understand when one stands on a fibreglass boat, which gets ice-slippery when wet. Were it not for my camouflage sheeting on *Water Horse,* I'd have been in myself several times already. The night following this incident, Ivor felt a sudden bump when his boat was at anchor, and shining his flashlight out the window thought he saw the 'Banana Boat' drifting off. Looking round however, to his surprise, he found it was in fact still tied up astern! By now, the other object has disappeared, and he wonders in consequence whether it could have been the Monster, swimming into the anchor cable underneath the boat, before surfacing nearby. I think it is possible, because he anchors in fifteen to twenty feet of water. I always back up for the night as shallow as I dare before hooking up (there is an art in doing this properly), but in shallow water one is just that much nearer the rocks, and if the wind changes a bad situation can develop.

Waking up to cast off in pitch black in rough water and pouring rain is not my idea of fun—nor is the journey across the bay in search of shelter. I've had to do this several times, with no more than a single glint of light a mile away far up on the mountainside to act as a beacon. It's lonely, and wet and dangerous and one longs to get back into one's bunk, and forget about boats and weather—but then the calm water welcomes you, and pretty soon the sun comes up again. Another day. Experience is the key to boating on the Ness—and what I call the 'belt and braces' technique. Never should you rely on one 'string' only—the term we amateurs have given to all ropes, sheets, cordage, lines and cables, regardless of type or thickness. It's all 'string' to us, though we do admit to 'best string' sometimes, or 'my longest string.' We've even got the professionals calling it string in some cases, which makes it a lot easier.

On August 26th I went down-loch and aboard *Prince Maddog,* the big research vessel, which I had first spotted from *Cizara* last year. Isn't it odd that on the same day that I logged the appearance of this strange vessel in the Ness (the 19th September 1968) the secretary of the 'Endeavour Society' of the University College of North Wales should write inviting me to lecture, and that when I got there I found they owned the vessel and had dinner with its chief scientific officer who had been aboard at the time. As I went alongside in *Water Horse* this summer the same Dr. Simpson invited me on board to take a look around.

The ship was anchored in 600 feet of water, and was involved in deep thermocline testing. It was, of course, packed with electronic gear, and had six scientists aboard as well as a crew of eight. She was in the 300-ton class, and while we were on deck a most curious trick of light occurred, which I tried to record on film. Somehow, the surface ripples were reflecting sunlight like the carrier wave of a radio beacon. It produced a unique effect, with waves of light flashing towards and past us at perhaps a thousand miles an hour—or appearing to. I have never seen anything quite like it.

We had supper aboard, served by a steward wearing a spotless white jacket. It transported me back over the years to my youth, and China, and meals served aboard coastal steamers. The thrum of the engines in the background, and the unmistakeable odours of a ship remain unalterable.

It made a pleasant interlude, and on the next day Ivor's boat and mine, the yellow submarine and the LNIB's lifeboat all became involved in a film sequence shot from the air by a helicopter, in the middle of the loch, opposite the castle.

It was great fun, and a bit of a circus stunt, with the downdraught from the 'copter kicking up the spray and the roar from its engines attracting throngs of tourists, who stopped their cars to watch. The submarine dived and resurfaced, with the boats whirling round it, then spinning off like sparks from a Catherine wheel. In the middle of this I ran out of fuel but managed to tank up again without spilling too much. Petrol mixed with oil is slippery stuff, most undesirable in the cockpit; and there is always the risk of fire. The next day our boats moved some fifteen miles down-loch to Invermoriston, with Ken Peterson and his camera-men aboard to shoot background film. They returned by road, so Ivor and I decided to anchor there for the night, because I had trouble with my engine and needed time to fix it. I tinkered about without much success, then found the main jet was blocked and cleared it. We decided in consequence to set off back to Urquhart Bay. It was dusk when we started, but our two boats finished the long journey together, lights glowing eerily as we rounded Castle Point, bucking through the rough water in the dark.

I had grown to accept the company of Ivor's boat, and its cheerful occupant, and when he and the Petersons left the scene of operations Urquhart Bay seemed a very empty place. I knew Ivor would be back later, but in the meantime I was alone on the water again at night.

The deserted shoreline at Loch Ness has a character which varies from place to place. It can be tranquil and secure, or black and foreboding, and exposed to every squall; and at one place in particular, a mile-long deserted stretch, I experienced a curious sensation of unease at night. This was so powerful I had to force myself to overcome it, and so real I mentioned it to people at the LNIB and also Mrs. Cary, who has a long experience of the Ness. In her younger days she fished a great deal, and knows every yard of water locally. She told me the same stretch of water had much the same effect on her, years before, and that when rowing home she would avoid fishing there. There was something 'unpleasant' about it. Even more peculiar—a few days later by chance she happened to read a paragraph from an old book called *Urquhart and Glenmoriston: Olden Times in a Highland Parish* by William Mackay LID, which had been lent to her. The following extract is from Chapter 21, 'Folklore in the Parish':

> According to tradition the Urquhart witches were, hundreds of years ago, the bearers of the stones for the walls of Urquhart Castle. These stones were brought from the districts of Caiplich and Abriachan, and the rock from which the wretched carriers got the first sight of the castle 'Cragan nam Mallachd'—The Rock of Curses. The great place of meeting of the Urquhart witches was 'An Clarsach'—The Harp, a rock on the shore of Loch Ness and within the bounds of the farm of Tychat. There they could be seen congregated on certain nights under the presidency of his Satanic Majesty, who sat on the edge of the rock and when not engrossed in more serious business, played to them on bagpipes and stringed instruments—which circumstance gave the rock its name…Their evil influence was exercised quietly and in secret and involved the object of their attentions in misfortune or even death.

The section of shoreline which interested us is below an estate some-times referred to as 'Cathouse' by the local people, but its official name is 'Tychat.'

I have yet to establish where the Harp Rock is, but it cannot be far from the spot where I was anchored.

On 10 September, Ivor and I played a part in another typical Ness experiment which had all the usual elements of drama tinged with farce. It had been decided to 'drop-test' Dan Taylor's one-man submarine in deep water, using *Lochalsh,* an old motorized barge on charter from Inverness. The idea was to lower the submarine's empty hull up and down three times to prove its integrity.

Fussy Hen, the orange lifeboat, first appeared towing the sub out of Urquhart Bay. She was packed with a crew of young men from the LNIB, mostly in frog-suits, who reminded me of the Keystone Cops.[5] She flew three flags, incomprehensible to all but the students of the International Signal code. They stood for 'submarine down' and, if nothing else, met the requirements of the Caledonian authority. In mid-loch the tow came to a dead stop awaiting *Lochalsh,* and Ivor and I came upon it in our boats, trailing dinghies. The temptation was too great and we circled them at full throttle, round and round, making Red Indian noises. It was one of those silly situations, and set the mood for what was to follow.

Lochalsh hove in sight, gradually making up to us. We tied up alongside, and in the hours that followed the two-ton midget subma-rine was cranked up and down to a depth of several hundred feet. I say 'cranked' because that is exactly the right expression. Incredibly the Keystone Cops had a great amount of cranking to do. It must have been the first time that anyone had ever wound a submarine up and down by hand.

While all this was going on, the whole collection of boats was drifting slowly down the middle of the loch, and opposite Urquhart Castle they caused a traffic jam, as visitors in hundreds peered open mouthed.

With the job done *Lochalsh* ploughed back to Inverness, leav-ing the submarine to hitch up a tow again, but during the test it

5 A fictional, incompetent police force who appeared in silent films of 1912–17 vintage. The group of bungling policemen appeared alongside the likes of Charlie Chaplin and Fatty Arbuckle.

had shipped water through the hatch and came perilously close to sinking. For a moment the situation was truly critical—saved by Ivor Newby, who abandoned his own boat, rolling backwards into the icy water in his frog-suit, to join the struggle to connect an airline. His boat was still motoring when I lassoed it. I think his quick action saved Viperfish, which was blown clear of water and de-ballasted on the spot by throwing rocks out through the conning tower, a most peculiar sight to watch.

Later that afternoon, I became involved in a film-shooting sequence for Scottish TV, who wanted to do a piece on *Water Horse* for a new programme called *One over the eight*. Stewart Henry, the disc jockey, was to put the questions, and as the sun came out brilliantly that afternoon, instead of working from shore we moved out on to the water of the bay. The Keystone Cops again—with cameramen and sound recordists, female directors and secretaries crowding aboard.

As you will probably remember the Independent Television News people did a long interview some weeks back, so I had no qualms. TV work is always interesting, and the people equally so.

Later that evening I met Bob Love, the American, who had arrived to start sonar operations in *Rangitea,* which has moved into the bay. Work started at once to fit her out with the mass of electronic and sonar gear, and I was soon invited aboard to look around.

Preparations for the sonar search are now extensive, and in a few days the loch will be 'pinging' loudly, and I must pay full attention to the surface, because I feel there is just a chance the noise may pop these creatures up out of fright—either that, or down, in the other direction to hide.

Operations became so involved and tense the last few days of the marathon I found myself spending 15 or 16 hours on watch each day, moving *Water Horse* about to places of advantage, and there was no time to write letters, or post them. There was barely time to go ashore for food and water. I knew the giant's hourglass was running out for me, and determined to put in a final effort. On 12 September, starting at daybreak, I became involved in another bizarre experiment in underwater acoustics, in company with the lifeboat and Ivor's craft. At intervals we hove to and dropped weighted light bulbs overboard, working to a signal code flagged between the boats. We knew these would collapse, or 'implode' at a depth of 600–700 feet, producing a noise like a pistol shot, which echoed through the loch. It all went nicely, but of the Monster there was no sign, so we turned about and motored back in rough water. I had a young American aboard as second camera man, and for a while I let him steer the boat, I stood up in the hatch for'ard, with head and shoulders poking out, watching the hiss and flow of the water rushing past. It has a mesmeric effect, but I was brought back to reality by the nose of *Water Horse* burying itself in a wave, and throwing the top of it straight into my face—about a bucketful of ice water. It deflected off me and over the helmsman, but there was something ridiculously funny about it, like slipping on a banana skin, and we both laughed.

One thing about boat work on the Ness is that it creates an extraordinary bonhomie amongst the monster hunters, a sort of excitement. It's like flying—I've seen this in people's faces, freedom and colour and light, all mixed up together—intoxicating. It makes up for all the physical effort and interminable lack of results, beyond the occasional scrap of film, or blip on a sonar screen. It balances the scales.

Within a day or two of this episode, Birmingham University's sonar on Temple Pier started to beam across the loch, and I was invited to view the sonar-scopes. They were confusing, but I was told that practice quickly enables one to pick out moving objects. Boats leave a broad band of dots, reflections of bubbles off the propeller, masking the screen; and for this reason I was asked not to cut across the bay, but rather go round the perimeter in *Water Horse*.

The sonar search was intensified when the Plessey team in *Jessie Ellen* put out one morning, and set up a loud underwater noise. One could even hear it on the surface, and the racket underwater was ear splitting, a high-pitched pinging sound. On two mornings this was augmented by rattles, towed along the bottom from either end of the loch by boats: *Scot II,* the converted tugboat from Inverness, and *Vigilant,* a pleasure cruiser from Fort Augustus. The rattles were of the type used by naval ships to mislead acoustic torpedoes homing in on them.

The whole operation was intended to drive whatever was underwater through the Birmingham sonar screen, where it would duly be recorded; but it did not have this effect, and because of the blanket coverage given on television news, this failure produced an anticlimax. But it did not, in my opinion prove, or disprove anything. I was there, on the spot, throughout these operations, watching from *Water Horse* and it struck me forcibly how local they were. The vast expanse of Loch Ness, stretching to east and west was unaffected.

This point was brought home to me shortly before my departure, when I left Urquhart Bay one jelly-calm morning, and idled down to Foyers, where a trawler skipper had reported seeing two monsters on the surface together the day before. As soon as I left the bay I realized how small it was in fact—like a children's playground, with all the activity and excitement—only a twentieth part of the surface water, the rest of which remained as empty and undisturbed as I had found it in *Cizara* the year before.

I anchored just off The Island, and on my way back, surfing in rough water, just a lonely dot, I realized how much I had become a water animal. There was no fear…With the passage of time the hunt drew inexorably to its conclusion. At first there had been so much time I scarcely thought about it, the days the weeks blending into each other. But now I was faced with the reality of failure. In ten years of relentless effort, and through eighteen expeditions I had not been able to improve on the film I shot in 1960. I had been close to the Beast on several occasions, of that I was certain, judging by the water disturbances which could not otherwise be explained, but these did not constitute proof. I wanted close-up film, but knew inwardly that I was not going to get it.

I had watched the seasons change from summer to autumn and one afternoon with little *Water Horse* peacefully at anchor close to the ancient castle ruin, a flutter of coloured leaves drifted down on to the water, like teardrops. It was a sombre moment for me.

I stood for a long time in thought, the ice cold of the water conducting through the hull, freezing my feet. Nature is so poignant, so inexpressibly beautiful on occasion that it holds a message for us, a form of visual poetry, rare fragments of which have been contained in verse. 'He leadeth me beside the still waters'—the words of the Old Testament poet. The hours went by, and as the evenings were closing in I withdrew to the shelter of the cabin.

Of one thing I was now absolutely certain. On an individual basis, the odds against a repeated sighting were so long that one could not count on defeating them. I had proved this. I might succeed in the next ten minutes, or perhaps not for another ten years. That was the measure of it—and as my whole private expedition effort hinged on the assumption that one man working efficiently, with the best equipment, for a long period, could probably beat the odds, the sooner I admitted this mistake the better.

But, as the Monsters were alive, and active, judging by my own experience during the summer, with wakes and disturbances, and sighting reports from people in boats who had been close enough not to be mistaken—if there was a solution at all it must lie in the technical approach; sort of communal effort put up by the LNIB and other scientific groups.

The morning of 21 September dawned clear, but the sunrise had a deep cherry hue to it, reminding me of the day on which I nearly lost *Cizara*. I took note of it because I have learned to read the sky like the Ancient Mariner, and at 12 o'clock I up-anchored and moved across the bay to bid farewell to the sonar and submarine people, the LNIB crews and all the other friends I had made. The sonar work was still in progress, and would continue in *Rangitea* for two or three weeks yet, so a lot depended on Bob Love's coming efforts.

I settled down to the long journey back to Inchnacardoch, the rain hissing in torrents, the water glassy, but I could not forget the ominous sunrise. Cherry-red meant trouble. Wind in all probability; and two hours later it arrived. Up till then the journey had been uneventful, with magnificent swaths of rain and mist shrouding the mountains, but I was numb with cold and, with several inches of water in the cockpit, had lost contact with my feet. I dropped anchor near Cherry Island, and went ashore to find Ivor waiting for me in his bright red sports car. He whisked me back to Urquhart Bay, where I picked up the Jaguar. I was too tired and cold to extricate the boat, but was determined to spend this last night—the eighty-second—aboard. In the meantime I visited the MacDonalds at Inchnacardoch Hotel, who kindly treated me to dinner. The warmth and comfort of the hotel was inviting but outside the massive stone building the wind was roaring through the trees, which bent and swayed together. It was more than a gale, and I realized the danger. Down the restricted cleft formed by the walls of the loch the wind could reach hurricane force, and any small boat that drifted out would be either sunk or pounded on the rocks.

I had to get back to *Water Horse*. The surface of Inchnacardoch Bay was protected but to reach the boat I would need to use the rubber dinghy. The wind and the noise were unnerving. I moved up wind, to allow for the drift, making an interception, then climbed on board, nosing *Water Horse* right up under the trees, almost into the mud. All round me in the bay I saw strange whitish blobs in the dark and, wondering if it was wreckage, turned on the blazing spotlight. The bay was full of wildfowl taking shelter. They showed no signs of alarm, which proves that in extraordinary natural circumstances animals lose their fear of man.

As I had become so much a water animal myself, I felt at home with them and huddled down in my bunk to sleep, but as so often happens on Loch Ness the wind dropped as suddenly as it started and I rose to a tranquil dawn. It hardly seemed possible; the whole experience was unreal, as though it had never happened, and yet all about lay the wreckage from the wind the night before, and ancient trees uprooted. The whole episode has taught me a lesson: that at Loch Ness no one can entirely master the environment.

Chapter 12

Shortening the Odds

Tim finished his long expedition at the same place he started eighty-two days earlier at Inchnacardoch Bay, coming full circle. In those days his experiences, diverse and many, were exceptional. It was his luck to be there and play a part in so many of the wonderful happenings during those summer months. Amazing discoveries happened almost daily. The submarine Pisces played an important role in many of the findings, with the sink holes in the loch bed and the loch's great depth previously unknown being of significant scientific importance. The unexplained sonar contact in mid water was, again, a morsel of intrigue, something to make one scratch one's head in wonderment as a forty-foot "something" doesn't just beetle off when approached—unless, of course, it's animate!

Another long and arduous drive south. Empty handed and melancholy at the lack of achieving his goal, Tim's downcast feelings were fuelled with a personal realization that the chance of capturing Nessie up close and personal on film was very slim indeed. It was a lottery, pure and simple. Perhaps it was your turn to win and perhaps not. In between time you just had to wait it out.

Waiting, or more accurately, hunting reactively was something Tim had done for the past nine years. It was a strategy made of sound logic: concentrate on an area known for sightings (not least his own incredible one) set one's self in a position with uninterrupted vision and top quality film equipment, give it enough time and, bingo, you have your result. Sounds simple and, on paper, there's no reason why it shouldn't be successful; but, as things were proving, nothing to do with Nessie was ever straightforward. Logic doesn't work with this girl, which in turn makes the whole mystery so much more compelling.

Tim arrived home thinner than when he left, suntanned, a bit weather-beaten and tired. The long expedition had taken its toll both physically and mentally and he needed time to recharge his batteries. He also desperately wanted to spend some time with his family. Adjustment was slow, as life down south was far removed from the one he'd just left; all the talk was about normal everyday family stuff. I was struggling with my schoolwork while both my sisters were excelling, Simon's first year in the Army had settled down and promotion was on the cards, and Wendy went from strength to strength in her career within the Citizens Advice Bureau—all a far cry from submarines, sonars, and monsters.

The great expedition of '69 lingered on a while longer after Tim left, with Bob Love and the crew aboard *Rangitea* scoring a sonar hit in late October. With all the media attention the summer's activities received, and disappointments at not having instant results, people were hungry for what they could get and Bob's findings were widely reported on the news. Bob's team filmed the sonar screen showing a bunch of white blobs on a black background, one of the blobs moved up and down a few times which was about as exciting as it got to the untrained eye. However, to the experts this blob was very interesting indeed; recorded in mid water between 210 feet and 540 feet down it moved around on a "looping path" for two minutes nineteen seconds. Bob and his team made no claims other than saying "in its motions it appeared to be animate."

However, no matter how many times the experts gave an interpretation for sonar blobs and squiggles—a large salmon showed up as a dot the size of a pin head while a "hit" could be an inch long—the results were only ever going to be accepted as that, an interpretation.

We sat together as a family in front of the TV, excited, watching the much-hyped news report and listening to Bob explain what it all meant. Dad, as always, was enthusiastic about the importance of the results and pleased for Bob and his crew to have a tangible reward for their efforts.

Tim understood and appreciated the importance sonar was playing in the overall picture; he also accepted it would only ever play a bit part in the puzzle, as dots, blips, and blobs don't go very far to convince an already sceptical scientific establishment about a large forty foot "something" living in the loch, no matter how many supporting charts showing otherwise were produced. No, to prove the monster's existence, without a shadow of doubt, it had to be with clear colour ciné film, surface photography, showing the animal in such a way that there was no room for doubt.

Later that autumn, once Tim had a chance to recoup and rethink, the LNIB Board of Directors placed an opportunity before him. Tim had a longtime connection with the LNIB and admired all the work David James and the other board members put in behind the scenes. He was asked to run the entire LNIB 1970 summer operation. It would be quite a task, one that would certainly curtail his personal watching time and take away the freedom of being the independent operator he so enjoyed. Flattered by the offer, Tim needed to consider how the marathon five-month expedition would affect not just him but also his family. The three months he'd just spent away from his wife, children, and home had been taxing enough, so he wanted to look at both sides of the coin before making a decision. He discussed it all with Wendy who, as ever, was supportive and practical, suggesting perhaps the family could spend some time at the LNIB's Achnahannet base camp during the school holidays. It was another one of Wendy's seemingly perfect solutions. Tim also weighed up the pros and cons of taking on the challenge, and before long realised the positives far outweighed the negatives. It was a wonderful opportunity to try a new aspect of his unique profession; plus, there was the added bonus of receiving a small honorarium for the position—meaning he might actually come out of the Nessie-hunting season financially even, which would be a first. He accepted the offer and with it the title of surface photography director.

With the burden of the decision lifted, it was, once again, full steam ahead with preparations for the coming hunt. Plans were drawn up and with an almost military-like focus Tim tackled the innumerable tasks of getting the project and the site ready for the inevitable influx of keen volunteers eager to chase Nessie.

Some years earlier, while attending the Farnborough Air Show,[1] Tim had witnessed a demonstration of the Wallis Autogyros. Intrigued, he set about finding out more information on the little craft and its capabilities. At first sight it looked an ideal solution for the shelved Oper-

---

1 Situated in Hampshire about thirty miles southwest of London, the airshow is an international biannual military and commercial aircraft trade fair. Its origins date back to the 1920s when the show was specifically for the RAF.

ation Albatross project. The original idea of Operation Albatross was to use a powered glider to fly over the loch and keep a vigil on the surface waters. If Nessie showed, the glider would shut off its engine and swoop down, taking pictures. The plan had some flaws: one was the need for a sizable runway and another was the lack of all round visibility. The autogyro scored high on both of these points. It was an open cockpit so the pilot had unrestricted vision, plus the little aircraft only needed the grass cut in a farmer's field and a hundred feet or so of smooth ground and it was airborne. The autogyro also had the added bonus of being kitted out to do reconnaissance flying, having a camera already fitted beneath the pilot seat. With the rotating wing it could throttle back and glide quietly down upon its prey.

A meeting was arranged and both Tim and fellow monster hunter Ivor Newby drove up to Ken Wallis' stately home in Norfolk. Ken, a retired RAF pilot of vast experience (he had flown twenty-four European wartime bombing missions, and was a true inventor and world record holder in aviation) was there to greet them and listen intently to Tim's scheme. It was an interesting proposal, and although Ken was rather incredulous toward Nessie he got caught up in Tim's net of enthusiasm and agreed to come to the loch in the summer and do some flying. The addition of Ken and his unique flying machine would just add to the colour of the coming summer's expedition. Another character was on board and with Ken came the anticipation and the excitement of a plan coming together, as Operation Albatross had been lingering for some years and now it was finally about to happen.

Everything going forward was new ground. Tim was under no illusion; a hugely challenging summer lay ahead. However, in spite of the challenges the odds looked good for a successful hunt. It was going to be a unique moment in time bringing together an assortment of people from all walks of life and from all corners of the globe to converge their efforts in a common goal, namely to answer the questions surrounding Loch Ness and its fabled inhabitants. The added bonus was Tim would get to share this amazing experience with his own family.

Chapter 13

Achnahannet Antics

On arrival in early May, Tim found a somewhat dilapidated Achnahannet site. A winter had passed and without any nurturing the place was looking really rather sorry for itself. The snow and storms of the season past had taken their toll on the array of green painted caravans: batteries were flat, equipment required repairs, and water had got into the kitchen. Everything, it seemed, needed attention. The ever reliable Rip Hepple, the LNI's (as it was now referred to) resident Mr. Fix It, and Tim set about making the place habitable and ready for the first crew who would arrive in the coming weeks. Sleeping quarters were aired, washed, and cleaned; the hole in the kitchen ceiling repaired; and a coat of paint on the surrounding fence did wonders to brighten up the place, and their moods, considerably.

The site was quite primitive. It had no electricity and only a meagre amount of running water (siphoned from a local spring), which meant no flushing loos. The accommodation was split by gender: the men bedded down in the "Black Hole," an aptly named caravan, and the ladies slept in the almost palatial by comparison "Ritz," a caravan of a similar vintage as the men's, only much tidier! Both accommodations were bunkhouse style where the volunteers would claim a spot by depositing their sleeping bag and kit on a first-come, first-served basis. The sleeping arrangements, although relatively comfortable and dry, didn't always work out. Snoring was a main complaint and lack of space and privacy were often blamed when, on occasion, crew members could stand it no more and would "slope off"[1] to The Lodge Hotel in Drum-

---

1   British version of "schlep off."

nadrochit, a small town about four miles away, to get a hot shower and a decent night's rest. These defections, however, were relatively few and far between as, for the most part, people accepted the closeness of their neighbours with good cheer and a healthy dollop of humour.

Once things got up and running, a twice monthly routine was quickly established; the new crews would arrive and be given an orientation plus the all-important camera drills. Every LNI member needed to know how to operate the equipment, for which Tim took responsibility. Once proficient with a camera they were then placed on a two-day schedule: one day on mobile watch, one day at base camp. Mobile watch consisted of spending the day at one of the watching sites dotted around the loch. Base camp duty meant tackling one of the innumerable tasks needing attention in and around the Achnahannet site. A roster was devised giving every crewmember a turn in each area. This became much talked about as folks would swap, barter, and even sell their shifts, some wanting to get out of kitchen duty whilst others tried to avoid the loneliness of a particular watching site. And then there was toilet duty—the daily task of emptying the toilet buckets from the two accommodation caravans. It was a simple process of collecting the buckets and taking them down the hill, through a gate, into a field from where it was another twenty yards to a cesspit where the buckets were emptied.

For some reason this wasn't the most popular of tasks, but as with so many things of an unpleasant nature, humour developed around it and The Great Elsan Race[2] was created. The race consisted of competitors being timed from the moment they entered the first caravan to collect the toilet bucket to returning both buckets empty and ready for use. The rules were limited to making sure the gate to the farmer's field was closed after passing though, in both directions, because one time it was left open and the neighbouring sheep decided the grass was greener on the other side of the fence and invaded.

The race became a much enjoyed diversion with times being posted for all to admire in the mess hall. The discussions would quickly turn to tactics and often the legitimacy of a particular person's time due to the method they used. However, as there were practically no rules at all, it was quite futile to dispute a winning run. Sadly, the race lost a lot of its

2 Elsan is a trade name for a portable chemical toilet commonly used in caravanning, boating and camping.

appeal when, on a particularly wet afternoon, a competitor went for a blistering time and took the hill far too fast for the conditions, losing his footing on the wet grass. The inevitable happened when the two buckets, one from each hand, shot into the air discharging the contents all over the sprawling competitor. The incident had far-reaching effects once the story got around—the numbers signing up to race and chance their arm at immortality dwindled considerably.

For the record, Rip Hepple is the reigning champion, recording a title-winning time of 2:19:05 seconds. Although he was applauded, it was noted (by those who knew, but didn't really care) Rip had far more practice than anyone else and so had an unfair advantage.

The PR centre also needed daily attendants. Flocks of tourists stopped in the small car park and paid a few shillings (it was still imperial currency) to wander through the rather cramped yet informative displays of Nessie-hunting history. For the most part the visitors showed great interest, asking endless questions, with the vast majority getting caught up in the drama of it all, often stating belief in the monster's existence. In addition to the entry fee, pamphlets, pictures and books were all for sale. The little centre did brisk business and applied any revenue generated toward running the site.

It was mid July when we—Wendy, Alex, Dawn, and I—joined the expedition. Tim had arranged a modern caravan for our use. It had a large window facing the loch, which helped to make the interior bright and breezy. It suited our needs admirably, and compared to either the Black Hole or the Ritz it was quite luxurious.

As new members we too had to go through the mandatory orientation process, which included learning how to use a camera. That was the part I particularly enjoyed. The main camera stood on a specially built viewing platform casting a cyclops eye over the vast expanses of water below. The massive 35mm lens, almost three feet in length, gave the whole thing the look of an anti aircraft gun, or a machine gun post, a sort of relic from the Second World War. It was all very impressive, and, once we mastered handling it all, very enjoyable too. The watching site was at the rear of one of the caravans, a few feet away from the entrance to the viewing platform. Whoever was on watch had to run down onto the platform and flick a switch situated halfway along on the left side to sound a bell alerting the rest of the site a target had been seen and was about to be filmed, then get to the mammoth camera and swing it around to point it toward the target, zero in using a side-mounted sight, and then hit the switch and start to film. A time of ten seconds from start point to the camera rolling was considered at the top end, whereas twelve seconds was the average, and anything less went into the "must try harder" category. While my ten-year-old legs would take me down onto the platform at great speed they just weren't long enough for me to see down either the side-mounted sight or through the viewfinder—a perplexing problem. As a member of the team, I needed to be on active duty and capable of performing this very important drill. After much consideration and discussion between senior expedition members, a special dispensation was granted and a small box appeared for me to stand on. It did the trick and enabled me to go on and post some pretty respectful times.

A small harbour was also being constructed below the Achnahannet site. Successive crews took it in turns to spend time adding to the building of the project by collecting and dropping rocks into place. The shoreline just below the LNI's HQ sloped out at a fairly shallow angle for about twenty feet before plummeting into the abyss; it was just enough for the two protruding walls of rock to provide a safe shelter for a couple of smallish boats. *Water Horse* fitted comfortably, along with the expedition workhorse boat, *Fussy Hen,* and even Ivor's larger *Kelpi* could just about squeeze alongside.

Construction was ongoing since the endless battering of the waves meant improvements and repairs were constantly needed. Day-trips were taken in *Fussy Hen* to scour the shoreline in search of anything that could be added to the harbour walls, and it became a source of bragging rights when something particularly useful was found. My role was primarily to help with building steps down the steep embankment. Lots of equipment was transported via the precariously steep path between HQ and the harbour, and foot slippages were commonplace. To prevent accidents, a series of steps was cut into the bank. My job was to fill up buckets with stones from the shore and place them on top of each step, thus providing grip. It was an important role that I took on with gusto, setting about the task at hand and challenging myself by setting targets of how many steps I could get done in a day. I very much enjoyed being part of the harbour team and occasionally would wiggle my way into a wetsuit—which was miles too big for me—and splash around in the shallows helping with the placement of rocks and stuff.

Being constantly surrounded by monster talk I admit to becoming a little nervous when descending to the loch side to replenish the buckets with stones, especially when there was no one else there—so I wouldn't hang around long!

LNI undertook countless experiments from their HQ on the loch; it was a hive of activity. One question they attempted to answer was whether or not the loch could provide enough food to support a group of large animals. Not knowing the physical makeup of Nessie, everything was speculative; however, with any animal of such size a reasonable quantity of food would be required to sustain a colony. It was a question that had been asked many times, so they undertook experiments to get an understanding of what could be a reasonable food source. Eel traps were laid off the Achnahannet shore, in about 200 feet of water, with the captured eels being studied back at HQ. The traps seemed to always produce a good catch, revealing the loch harboured a very healthy eel population. On the occasions when hooks and bait were laid often the whole lot would be taken, apparently indicating the presence of some very large eels.

The weather that summer was appalling: gray, cold, and wet day after day. The wind whipped up the surface of the loch into a boiling mess of waves and whitecaps making surface monster hunting quite pointless. It was more like autumn than midsummer. Ken and his autogyro made flights in-between bouts of bad weather, and the sonar teams continued with their various experiments in spite of the inclement conditions. The show must go on, and as many of the folks had travelled many miles to be there, good humour prevailed. People buckled down and got on with what they could.

Tim had scheduled a side trip for the family and Ken to Loch Morar, a smaller loch on the west coast of Scotland, about eighty miles from Loch Ness. It was also reputed to be a place where strange animals were seen periodically. Morar was quite different from Loch Ness in that it wasn't a tourist destination, it wasn't looped by a road, and the locals liked to keep the stories of "Morag"—their Nessie—quiet. All this changed, though, in summer 1969 when two friends out fishing— quite a regular pastime for them—on a fine day in August reported an unusual experience. The water was flat calm, a nice day to go fishing. The following is their account as reported firsthand by one of the two witnesses, Duncan McDonell:

My friend and I were travelling down Loch Morar in our boat, we were moving quite slowly at the time. I was preparing a cup of tea when I heard a splash or disturbance in the water astern of us. I looked up and about twenty yards behind us this creature was coming directly after us in our wake. It only took a matter of seconds to catch up to us; it grazed the side of our boat. I am quite certain this was unintentional. When it struck the boat it seemed to come to a halt or at least slow down. I grabbed the oar and attempted to fend it off, one of my fears being if it got under the boat it might capsize it. I struck it with the oar quite hard breaking the oar in doing so, then I heard a report at my side, my friend had fired a shot. The noise seemed to scare it and it dived into the depths. After our ordeal we went ashore and this is when I realised the dangerous position we had been in, it really was very nerve-racking…

The news media were all over it and sensational stories flew about how the creature had bitten the oar in two, and that the two men were trying to cash in on the growing popularity of Nessie and bolster the area's own tourist industry. However, both were simply long distance lorry drivers who enjoyed fishing quietly at weekends on the loch, neither were in search of fame and certainly weren't pleased with the attention they received in the weeks following the news stories.

The LNI, and separately Tim, followed up by doing some research into both the reported incident and other sighting reports and local stories of Morag. Their respective findings gave reason to believe there was possibly more to it than just a good yarn and that it was worthy of a little more investigation.

Chapter 14

The Clear Waters of Loch Morar

We left a dank, windswept Loch Ness mid August in a small convoy, the big Jag in the lead, stuffed with the Dinsdales' equipment and towing *Water Horse,* followed by Ken Wallis in his little Austin Mini and the autogyro, and bringing up the rear was one of the LNI's green expedition vans with the Moo-scow firmly strapped to the roof. As we drove along we became conscious of passengers in passing cars pointing and staring at us. Motoring through the west coast town of Fort William, pedestrians would stop, nudge their neighbours, and point at our unique little road train, with Ken's flying machine getting the lion's share of attention.

Past Fort William we branched off north into what can only be described as a land of immense beauty and dramatic proportions. The west coast of Scotland is a wild place with craggy mountainsides and heather-clad moors littered with countless burns (streams) and rivers meandering their way toward the sea. The road quickly became a twisting single track with the occasional lay-by and numerous small, stone, single-arch bridges spanning the many waterways.

As we motored on we came to the impressive Glenfinnan Monument at the end of Loch nan Uamh. It marks the spot where, in 1745, Bonnie Prince Charlie of the exiled House of Stuart was met by clansmen prepared to battle the English for both the Scottish and English thrones and replace the sitting monarch George II, something they almost succeeded in doing.

As we got closer to Mallaig, the fishing port just a few miles from Loch Morar, the weather brightened and, before our eyes, the sun finally came out. For the first time in weeks we saw blue sky—a good omen for the coming expedition. Rounding a corner in the ever-narrowing road we got our first glimpse of Morar, the deepest body of water in the British Isles, which had quite a different look to the fiord-like scenery surrounding Loch Ness. With its islands at the westernmost end and treeless mountain vistas, the place felt very different from the landscape we'd left that morning.

Our plan was to stay at an old converted croft house about a mile along the shoreline. Before we could reach it the vehicles had to negotiate the road, which swept up steeply ahead of us at an almost impossible angle and passed over a bluff. Tim was duly worried about the Jag not making it while towing *Water Horse* and so opted to run the boat into the loch right there and then. There was a convenient launch ramp, and with Ken, my dad and me (I'm not really sure how much help I actually was) we got her safely on the water. Due to Tim's busy schedule at the LNI HQ, this was the first time *Water Horse* had hit the water all summer. Keeping with the tradition of adding a new piece of very important equipment before each season's hunt, this year the purchase was a powerful 40-horsepower outboard engine for *Water Horse,* which was pretty close to her limit.

After our first night at the B&B we launched Moo-scow and rowed over to the mooring where *Water Horse* had spent the night. Ken and Tim struggled with the new engine but it eventually fired up. Slipping the mooring, we slowly headed off out of the shallow bay toward the open waters of the loch. Just as Tim was about open the throttle the engine died. After much huffing and puffing the brand new Mercury simply wouldn't restart. Tim and Ken were both engineers but that didn't seem to help, as no matter what they pulled or pushed the darn thing just wouldn't play. It was about this time when the wind picked up ever so slightly, just enough to start moving *Water Horse* and her contents toward some rather ugly and sharp looking rocks on the shoreline. At first no one took much notice, but as the rocks started to grow in size so did the concern of the passengers, and, quite quickly, all aboard realised a collision was a very real possibility. Taking the initiative, I suggested I could tow *Water Horse* by attaching a rope to the little Avon inflatable Moo-scow and row away from the danger. At that point there really wasn't much option and so Dawn and I jumped in and I started rowing as hard as I could. It worked: by planting both oars in the water and standing up to put as much leverage as I could we slowly started to move away from the shore and the danger.

I was credited with saving the day, and once the engine was finally running again I was given the honour of taking the wheel. Under the watchful eye of my dad I opened up the throttle to give *Water Horse* her head with the new outboard now running perfectly. It was an exhilarating experience as she took off with her nose rising slightly out of the water and then settling to skim across the perfectly flat calm surface in a sensation not unlike flying. After the initial hiccup the test run was a success and the expedition, and monster hunting, was ready to get underway.

Tim met with a small group of enthusiasts calling themselves the Loch Morar Survey (LMS), which consisted of a bunch of young students, mostly marine zoologist and biologist undergrads from the University of London. They were undertaking a survey of Morar trying to find out if the place could sustain a colony of sizable predators. They also had watching sites, equipped with long-lens cameras, dotted around the shoreline. One of the more spectacular sites was perched high on a hillside, some 300 feet above the water, overlooking the west end of the loch and the cluster of islands. There would be no escaping the camera's eye if Morag decided to show in that vicinity.

The weather continued to be glorious and the flying conditions near perfect. Ken took every opportunity to get airborne; flying numerous sorties, he repeatedly went up and down the loch constantly scanning the crystal clear waters below.

On one such mission Ken came across *Water Horse* motoring sedately along between the islands. Dad, Dawn, and myself were on board enjoying a pleasant afternoon exploring when out of the blue—or should I say the sun—Ken attacked. We heard the autogyro coming but it was too late; Ken already had us in his sights. Before we knew it he came swooping over our heads, skimming us and *Water Horse* in what seemed just feet above us before pulling up sharply and banking away, eyeing us up for his next pass. Dad realized immediately what was afoot and opened up the throttles; *Water Horse* sprang to life and took off across the loch, ready to do her own "flying." This time he was prepared for Ken as he came in for the kill. Dad banked the boat hard to the right making Ken overshoot and completely miss his target. The cat-and-mouse game continued with Dad using the islands as cover to dodge and weave away from the attacks while Ken mixed it up by varying his strike tactics. It was exhilarating stuff that made us all feel like we were a part of a 007 movie—which of course Ken had been a few years earlier. He finally called off the attack and flew away waving a cheery farewell. Grinning from ear to ear, we set for home.

At dinner that evening we recounted the story amidst much excitement and thrill. Ken said at first he'd decided to use us for camera practice, but when Tim responded the chase was on. Listening to all our conversation and laughter was the very quietly spoken owner of the B&B, Mrs. Parks. A reluctant storyteller, she slowly joined in our chatter, explaining the different names of the islands and some local history. Before too long the topic turned to the monster. Mrs. Parks was the sister of one of the two men who had had the (widely reported) close encounter with Morag the year before. She had met up with them when they returned from that eventful fishing trip, as their boat was moored on the shore just below her croft house. She spoke about their fear and unnerve, and how before the incident both men had scoffed at any mention of the monster as nothing more than a myth. Now, however, with the story leaked to the press, their reaction was quite different; they were extremely annoyed, as neither had any desire to comment on their experience, realising how improbable it sounded. They just wanted it all to go away. Their annoyance doubled when the press started to exaggerate the already incredible incident by talking about the monster "biting the oar in half," which wasn't the case at all. When the animal brushed the side of their small fishing boat it tipped, causing the boat to roll and knock over the open-flamed stove in the cabin. It was alight at the time, boiling some water for tea, and both men feared it would set the boat on fire. William Simpson, McDonell's fishing companion, went below to sort it out. McDonell, left in the cockpit alone and fearful of being capsized, grabbed an oar and used it to try to leverage the boat away from the creature; it was in doing this that the oar broke. Simpson came out of the cabin with a .22 rifle and let off a shot over the creature's head, the crack of which seemed to scare the animal and it immediately submerged.

It was an amazing story and told with such sincerity that it would be hard to doubt Mrs. Parks—and anyway, why lie? What on earth would two lorry driving fishing friends gain by making up such an improbable yarn, especially knowing the ridicule they would surely face?

Dawn and I went to bed, not for the first or last time, with stories and images of monsters running through our heads, both happy to be sleeping ashore in the safety of a warm comfortable house without the danger of being tipped out of our bunks by some passing prehistoric beast.

The following days continued to be kind, with the weather staying fine. We toured the entire loch, and at one exciting moment thought we had a sighting of a large back; but it turned out to be a rock sitting in the water just off the distant shoreline. Dad made some good contacts with the LMS folks and enjoyed watching, and assisting, their efforts. The trip was deemed a success in that much was learned about the loch and its inhabitants, with one of the more interesting finds being the incredible clarity of the water. In Loch Ness the opposite is true. The water in the Ness is stained with peat particles transported into the loch via hundreds of streams and rivers that drain the surrounding mountains. The peat content vastly reduces visibility, giving the water the colour of tea. Indeed once divers pass below thirty or so feet they start to lose all vision and direction. Even with strong lights the ability to see is limited. Morar's water, in contrast, has no such limitation, a fact that would be of interest with expeditions to come.

Chapter 15

"What's that?"

Leaving the west coast after seven glorious days of sunshine, the mini expedition returned to find a much-disgruntled LNI camp. They had continued to experience a Highland summer at its worst. The bad weather pattern hadn't shifted and the barometer continued to show low. The returning group was met with some envy, as the suntans proved the stories of blue skies and warmth weren't fibs; the cheery talk of flat calm days, sunshine and carefree monster hunting was a little hard to take after the endless rain and wind everyone at the Achnahannet site had been enduring.

The "summer that never was" brightened one August afternoon when Tim, aboard *Water Horse* with son Simon and fellow independent monster hunter Murray Barber, were helping out with experiments in Urquhart Bay. The shout went up, "What's that?!" Simon had spotted an object moving through the water across the bay. Tim and Murray both looked up and saw what they later described as "looking like an eight-foot-tall telegraph pole streaking through the water." Tim immediately dove below into the cabin to grab his Beaulieu camera (the Cyclops Rig wasn't set up due to the busyness of operations *Water Horse* had been involved with). In the fleeting moments between first seeing the object and getting hold of the camera the monster was gone. It all happened in a flash, no time for filming or taking pictures just an adrenalin rush and the realisation that Nessie had been close, very close, but the elusive lady managed to get away yet again without waiting for her camera call.

The near miss added some spice to the otherwise wet and, so far, fruitless summer. In spite of the miserable weather, each passing crew enjoyed their time at the loch. Watch duty continued, as did the experiments and, of course, the light-hearted relaxation in the evenings at the Lodge Hotel where folk songs were sung and bagpipes played at the weekly *ceilidh*.[1] These evenings always helped to brighten any rain-dampened spirits.

Besides the lighter side of things, the intent and purpose of explaining the mystery of the loch was as strong as ever. The steady stream of underwater results had intrigued people who had knowledge of sonar. Toward the end of the summer, a group from The Academy of Applied Science, out of the US, arrived at the loch with their own highly complex side scan sonar unit. Dr. Martin Klein, the designer, along with Dr. Robert (Bob) Rines and Ike Blondel set up a number of experiments starting with the laying of bait. A concoction of scents and hormones known to attract aquatic animals was dropped in to the loch at various places. At the same time, the Academy team set up their sonar equipment in Urquhart Bay where they recorded a sizable target passing through the sonar beam. Encouraged with the early results, the next step was to tow the side scan sonar behind a boat along sections of the loch to see if they could pick up any more targets, or intruders as they liked to call them.

Tim offered the Americans the use of *Water Horse,* which they duly accepted, and with that began friendships that would last a lifetime and would take them all on an amazing ride of discovery and fascination at the ever deepening mystery.

The overburdened *Water Horse* motored slowly along, stuffed full of electronic equipment designed to interpret the sound waves being sent by the "towfish," as Marty had named the side scan sonar unit, 300 feet below. Great reams of chart paper were being spewed out with lines and shaded areas all quite impossible to understand unless explained by an expert. The results after two days of towing showed that, on a number of occasions, large intruders had been recorded, showing up on the chart as a large black blob compared with fish, which showed as a pencil dot. Marty's equipment also picked up the contours of the loch wall and for the first time revealed large undercuts and underwater caverns. It was all very interesting stuff and yet more fuel for the fire of the mystery.

Marty and his towfish were booked to do experiments off the coast of Norway and so, reluctantly, he left Tim and the Academy team behind to continue their adventures without him, but vowed to return.

---

1 Pronounced "kay-lee," ceilidh is the Gaelic word for a social gathering usually involving playing traditional musical instruments and singing folk songs.

Marty, like so many before, was another who had been gripped by the mystique of Nessie.

Ike and Tim, inspired by the sonar results, continued their experiments, this time using hydrophones, and were to have yet another very unusual experience while doing so. Motoring slowly under the Horseshoe, the area where they scored a hit with the sonar, Ike lowered a hydrophone into 600 feet of water. While on its descent, at about 200 feet down, the hydrophone hit something in mid water, stopping its downward travel. The hydrophone had come into contact with something solid. As it scraped along the object, loud rasping sounds could be heard coming from the speaker in the cabin on *Water Horse*. Rip had joined the group and the men stopped and looked at each other. They all knew there wasn't anything below them as they were in deep, open water; for a split second they questioned what was happening. After just moments the sound ceased as suddenly as it had begun and the hydrophone was once again falling freely down into the depths of the loch.

Amazed, they tried to rationalize what the hydrophone had just come into contact with. The thought of a sunken tree was the only reasonable explanation—other than making contact with the monster—they could conceive. A waterlogged tree trunk at 200 feet down wasn't all that plausible, yet hitting the monster wasn't either, so they just had to chalk it up to another one of those inexplicable incidents that seem to happen with remarkable regularity when chasing Nessie.

The Academy took away the data and spent much time sifting through it all and putting together a straightforward, no-nonsense report on their findings, which concluded as follows:

In summary our brief side scan sonar tests in Loch Ness in 1970 produced three important discoveries.
- There are large moving objects in the Loch.
- There is abundant fish life in the loch which could support a large creature.
- There are large ridges in the steep walls of the loch which could conceivably harbour large creatures.

It was a good way to end the long season's chase. It constituted more circumstantial evidence than before and, although not at all conclusive, it was enough to give the various interested parties and all those involved something to ponder over the coming winter months.

Chapter 16

American Enthusiasm

Tim had been chasing Nessie full time since 1968 and the routine was now a familiar one. Once the summer expeditions were over it was time to concentrate on paying the bills and preparing for the next round of battle. Paying the bills, an ever-present concern, was achieved by fulfilling numerous speaking engagements. The autumn and winter months were the best time of the year for this, so shortly after returning from Scotland Tim would be busy traveling the length and breadth of the country talking to all sorts of groups, schools, colleges, and universities. Over many years of public speaking Tim had honed his delivery and had become skilled at altering the content to suit each audience. For universities there would be a high level of technical content and data, for schools the emphasis would linger on the mystery of it, and for church and social groups the human element would feature strongly. And it seemed to work, judging by the number of repeat requests. In fact, the vast majority of his engagements came via recommendations or requests for repeats in which he was invited back to the same club or school so different members or pupils could experience the subject.

It was just as well his talks were liked because one thing Tim wasn't great at was self-promotion. It is hard to believe in this day and age, but back then there was no website, no sponsorship deals, no leaflets, not even a newsletter. In fact it wasn't until near the end of his life in the mid 1980s when Tim finally started sending out a promotional mailshot. He also charged very little in comparison to others on the talking circuit, preferring to give, as he saw it, a fair price for a fair service. Today that would be seen as a naive approach. He certainly should have been charging what the market would bear, and, in fact, with his level of fame (I use that word not in today's context of tabloid, TV, or music stars, but as described in the dictionary as "the condition of being well known," and for the subject of Loch Ness and its monster he certainly was that) he was charging far under his market value. Tim would always argue he didn't want to put folks off by being too pricey—killing the golden goose as he put it—plus the more people he could speak to about the monster the more acceptance the subject would get. He found that presenting the evidence, much of which was new to audiences, in a calm and professional manner greatly swayed opinions, and by increasing people's knowledge much of the old prejudices against the monster (such as the attitude that the stories about Nessie were nothing more than a myth, joke, or hoax and those who chased it were slightly barmy) would transform into an understanding of the serious work being undertaken in the pursuit of the truth.

Shortly after taking his film in 1960 Tim stated, "I would not seek wealth for its own sake, and if I made anything out of the film, or from articles about it, the proceeds would go first of all towards proper equipment, and the cost of future expeditions," and that remained so. After a decade of committed research and searches at the loch, Tim had written two books, both of which were well received. Unfortunately, the revenues didn't bring in nearly enough to make his financial life easy, so when, in spring 1971, the Academy of Applied Science invited Tim to give a series of talks on the east coast of the US, he readily accepted.

It was a successful couple of weeks meeting a variety of people whose attitude Tim found to be quite contrary to much of that of the British establishment where Nessie was still often seen as a joke or embarrassing to be connected with. The enthusiasm and energy of the Americans was a refreshing change and it helped to refuel Tim and ready him for the challenge of the long summer ahead. He had agreed to take on the running of the LNI for another year, only this time his responsibilities would be increased, as he would also be coordinating the underwater side of things as well. It was actually an exciting prospect as the Academy was preparing a new underwater initiative where cameras, developed at MIT in Boston specifically for the task, would be linked with a powerful strobe light suspended below the surface of the loch. Ike Blonder of Blonder Tongue Laboratories was continuing his hydrophone work and developing a playback system where any sonar noises recorded would immediately be played back to the source, and if that source was the monster then perhaps a series of back and forth communications could be established with Nessie talking to herself. Ike had recorded a sound pulse while in the area of the Horseshoe during the summer of 1970 which showed a definite pattern, and, according to experts, could have been produced by an animal—rather like the way a dolphin uses its own sonar to locate food. It was an interesting concept and one that would just add to the variety and colour of the summer's experiments.

Chapter 17

Brock's Brush With The Beastie

Back at the LNI's Achnahannet site, preparations for the coming season included the building of a new visitors' centre. The old caravan used in previous years just wasn't up to the job of supporting another round of thousands of visitors, so a much larger and sturdier structure replaced it. It was a far more practical design with separate entrance and exit doors, which might sound logical but as the LNI had merely stumbled upon the visitors' centre concept some years earlier thanks to tourists stopping and asking questions, the new purpose-built centre gave the whole thing a far greater air of professionalism.

The additions to the site hadn't ended there. Rip spent many weeks building and plumbing a small outhouse, which, sadly, put an end to the Great Elsan Race. The upside was that the little building now contained a pair of long awaited flushing toilets. It was luxury of a magnitude that expedition members from years past would have only ever dreamt about. The clean little washroom also had a sink boasting both hot and cold running water. This addition, without question, made Rip the HQ's most popular person of the season.

1971 got underway with the pattern of crews coming and going much the same as the year before. Having greatly enjoyed our experience in 1970, all the Dinsdales (Simon was on leave from the Army for a couple of weeks) joined the expedition again in mid July, and like old hands, quickly settled into the everyday running of the place. Tim had purchased the derelict lifeboat *Narwhal* from the Abbey sailing club for the LNI. It was the boat he had tied up alongside in *Cizara*. *Fussy Hen* couldn't keep up with all the water work going on, and anyway she was just too small to carry the vast array of equipment many of the experiments being undertaken required. The old lifeboat was in quite a bad way and needed a considerable amount of work to make her seaworthy again; she replaced the harbour as the favoured project. Once the hull had been scraped and repainted she was moored in Inchnacardoch Bay where a full refurbishment was undertaken. A cabin big enough to sleep a small crew was built, a transom with an outboard motor was added, and before long the old wreck, sporting a new paint job, was looking very much a part of the LNI's expedition force.

It was on a trip to *Narwhal's* refurbishment site when I took one for the team. Helping to launch *Water Horse,* I had taken control of the large, rusty, old hand-cranked winch. The boat was all but in the water and everything was going smoothly when the handle suddenly slipped out of my hand. With the weight of *Water Horse* on the end of it, it naturally flew around at great speed coming back to crack me square on the forehead, knocking me "basic over apex" (off my feet and up-side-down). I landed flat out on the ground, literally seeing stars. All those around came running over and showed huge concern, not the least of which was demonstrated by my mother. I was determined to show I wasn't injured. In fact it was my pride that was hurt more than my forehead despite the big egg-shaped lump forming. In the days that followed I wore the lump as rather a badge of honour.

As *Narwhal* neared completion, moorage needed to be laid for her in Urquhart Bay where she would be based and used by the group from the Academy. A pleasant chap named Brock Badger was on crew at the LNI and offered to do the diving required to float a large block of concrete—which would act as the anchor—into place. After lots of pulling and pushing and attaching numerous flotation buoys the anchor was towed into the lock, and with the help of Brock, who was in the water, dropped into about twenty feet of water at just about the exact spot required. It was a job well done, and as the shore crew were packing things up Brock was finishing off in the water. Watched by my brother Simon, Brock suddenly burst to the surface in a boil of foam and bubbles and swam at full speed to the shore. Nothing was said at the time but in the days that followed Brock just wasn't himself. Normally a cheerful chap with a great big grin and always keen to help, he had become rather reserved and quite introspective. Eventually Tim, concerned at the way he had been over the previous days, questioned if all was okay. Brock admitted to having seen something at the extremities of his vision, which at that depth would have been around thirty feet maximum. He talked about an object, big enough to fill his vision, moving from his left to right. It was definitely moving, and at that point he didn't hang around to ask for an autograph!

This report caused quite a stir at the LNI, as Brock was a very well liked and trusted four-season expedition veteran. He was definitely not known for exaggeration; in fact, far from it. He would more likely underplay a situation rather than overstate it. He had dived many times in the loch and so was no stranger to the murky waters and the possible light reflections that could sometimes happen; so when Brock quietly announced his diving days in the loch were over, folks who knew him took notice.

A year earlier two seasoned, professional divers also had a very strange experience while diving on a wreck site, again in Urquhart Bay. After losing contact with the vessel in the inky blackness the two men linked arms and started to ascend. On their slow rise to the surface their flippers came into contact with something solid. They both felt it but were at a loss to know what it was they had touched. On reaching the surface they were quick to exit the water and board the waiting support boat. Their contact coincided with a large sonar intruder entering Urquhart Bay, recorded by the Academy. Later the divers suggested they may have touched the mast of the sunken ship leaning at an angle, but this was only a guess, as they knew they were in deep, open water and logic says there's nothing there. Still, whatever it was the two hardened professionals had come in contact with unnerved them sufficiently; in the days that followed they requested the bait bags laid by the Academy either be removed or placed at some distance from their diving site.

Reports and incidents like these were welcome additions to the variety of surface sightings; it added another dimension to it all. The underwater route was yielding more results than surface watching, yet, as with eyewitness accounts, it was all still very subjective and open to both interpretation and criticism. The evidence still needed to be stronger, conclusive. The LNI, the team from the Academy, and Tim were all working toward the single goal of finding the truth. Tim knew what he had seen and filmed wasn't a boat or anything that could be explained rationally. After all, he watched it swimming and submerging, so unless there had been a very unusual submarine operating in the loch that day (which there wasn't) the only conclusion was he'd seen an aquatic animal of some unknown description.

Brock's report was an important addition to the ever-growing mountain of evidence. That it was just another eyewitness account and thus open to opinion didn't matter to Tim as he trusted Brock and was convinced he was telling the truth. Brock had recounted plainly what he'd experienced. He never claimed to have seen Nessie, just something quite large moving in front of him. The debate surrounding Brock's close encounter filled the mess hall at Achnahannet. The Americans joined the discussions posing the interesting question: Would the monster be in such shallow waters? Tim noted that the point where Brock had seen the object would have been very close to the edge of the shelf where the bottom plummets vertically from thirty to hundreds of feet of water, and, in fact, the monster could have been patrolling the shelf but still have been in deep water. It was an interesting concept and one that wasn't missed by the group from the Academy. They had been working on an idea of positioning a camera with a strobe light along the shelf; the plan was to take interval pictures, hopefully capturing the animal if it swam past.

A rig was set up and amid much excitement the expensive and unique equipment anchored into position on the shelf close to one of the many rivers feeding the loch. A couple of orange buoys floating on the surface marked the position. The whole set-up was left for the night, with the strobe lights merrily flashing away and the camera clicking in synch. Returning to the site the next day the Americans, and everyone else for that matter, were aghast to find the entire operation gone. No buoys, no flashing lights nothing; the whole thing had just disappeared. It was a disaster; a year's worth of technical development along with expensive equipment had just vanished. Grappling hooks were thrown from boats in the hope of snaring the rig, which perhaps had slipped into deeper water. Divers were called in and everyone spent the day peering over the side of boats in the forlorn hope of seeing the strobe light flashing somewhere below.

Theories were bandied about regarding what could have happened to it. Some suggested it had been stolen, others thought it might have got tangled up with the monster and dragged away, but the most likely was it was cut adrift by salmon poachers. The poachers would always operate at night, trawling the areas around rivers, hooking into the feeding fish. The rig was positioned directly where they would have been fishing, so now it was most likely the equipment was sitting at the bottom of the loch somewhere. Dejected and very annoyed, the party broke up and the Americans headed back to their lodgings while Tim made his way to the LNI HQ. Rip, who had been in attendance working from *Fussy Hen,* steered the little boat out of Urquhart Bay and toward the Achnahannet harbour a mile or two away. Motoring along he spotted a buoy out in the middle of the loch. Knowing it was deep water and very unusual indeed for a buoy to be positioned there, he went over to investigate. As he approached, he throttled back and then could immediately see the strobe light flashing away 10 feet or so below the water. He grappled with one of the buoys and secured it to the boat, then he swung around and slowly headed back toward Temple Pier where the Americans were operating. The recovery of the rig was a tremendous relief and boost to morale as so much effort had gone into the design, and big hopes lay in the success of the strobe and camera combination.

The rig was eventually repositioned, but this time not near a river mouth, and spent many undisturbed days flashing away in Urquhart Bay. It also had a spell in Loch Morar where the clear water was a huge bonus, giving the underwater cameras a range of sixty-plus feet. Alas, once all the film had been developed and analysed nothing more interesting than a few startled looking salmon had been captured on film. The lack of results did nothing to dampen Bob's and the rest of his team's spirits; quite the contrary in fact. They went back to the States full of ideas of what needed to be improved for the following season and set about making it happen.

Throughout the summer there had been a steady flow of people coming into the Achnahannet centre with monster stories of their own. Each was dealt with in an open-minded way. Folks would be asked to fill out a sighting report form, and Tim, whenever possible, would record each individual's account on a small cassette tape player. One afternoon we got to hear of some large three-toed footprints found on the south shore. Sceptical but intrigued, a small group from the LNI motored around the loch to take a look at the tracks. A young biologist was on crew at the time and so he packed along a bunch of sampling fluids and some plaster cast to take a print of the tracks. Mindful of the hippopotamus foot hoax back in the thirties, the group arrived and were shown a small beach area where there were about four or five prints, a couple of which were very clear indeed. Samples were taken and the plaster cast mixed and poured into the most defined of the footprints. As the plaster was hardening I began to get bored with the whole procedure and started skipping stones across the water, a pastime I particularly enjoyed. Rip came up and noticed an unusual shaped stick bobbing around at the water's edge close to me. He picked it up and immediately started using it to try to reproduce exact copies of the tracks. It worked; we'd been duped. Someone had obviously used the stick to make the tracks and then threw it into the loch, only for the wind to blow it directly back ashore. We all returned to our vehicles in relatively good humour, but again a little disappointed.

Early in September Tim was filling out a sighting report of his own, when, for only the third time in eleven years, he got a fleeting glimpse of the monster:

> I was motoring *Water Horse* through rough water towards Foyers Point, to try out a new baiting technique. It was mid afternoon and stormy and the surface noise from the hydro-works at Foyers was loud enough. Out of the corner of my eye I spotted a black snake-like object rear above the water; it stayed erect for a moment, then it was quickly withdrawn. By now I was staring full faced, spellbound, toward the object—unbelievingly. Surely it couldn't be the monster. After so many fruitless years of searching, my mind simply rejected the possibility.
>
> I continued to stare, only to see it break surface again, then go down in a bail of white foam. It was no more than 200 yards distance—I could judge by the size of the seagulls.
>
> Too late I throttled back and crept forward over the loch towards Foyers Point. There were five cameras in front of me. Because of my rooted attention to the place where the neck had broken surface I ran the boat aground. Fumbling angrily with throttles, I poled off and dropped anchor, throwing the bait bag over the side. It was clear unless I could get a hold of myself I would never film a thing. I put on my life jacket...
>
> The object I had seen was between 4 and 6 feet out of the water. It was mobile, muscular—and for a time almost looked exactly

similar to the "Surgeon's Photograph."[1] The one difference was in the extremity, which had no visible features; it appeared rounded like the end of a worm. It was suggested afterwards to me the object might have been the tail, but I rejected this. If it was a tail then it was moving backwards!…

This encounter shocked Tim. He realised after all the years of chasing the darn thing he'd got rather complacent. Although it was a very brief sighting, no matter how quickly he'd reacted there wouldn't have been a chance of shooting any film. Even a still photo would have been out of the question, as the monster had gone almost as quickly as he'd first seen it. The experience gave Tim both hope and cause for concern; hope in that finally he had seen the thing again, and concern in that he wasn't ready for it. In the early years he was always alert, ever vigilant. However, as time passed and each expedition faded into memories without so much as a sniff of the monster, so his ease with being at the loch had grown. The place was his second home, and he had become so familiar with the surroundings that his primary purpose for being there had been diluted with all the wonderful—yet distracting—activities he was now involved with.

Tim resolved to return to a state of constant readiness. He realised that due to the shock element of seeing the creature, camera drills couldn't always be relied upon, and he set about developing a technique he christened "shooting from the hip." It was simple and involved nothing more than grabbing a camera and literally filming from the moment it was pointing in the general direction of the target. Of course, the Cyclops Rig stayed, as always, but the numbers of loose cameras around the cabin increased. Tim's mantra had always been to have a camera within an arm's length, and for the most part he had kept to it, but with the number of distractions going on it had become an easy thing to let slip. He would consciously redouble his efforts to be in a continual state of expectation, meaning a sighting could happen at anytime!

The season ran its course with numerous faces coming and going from the Achnahannet site. The newly built exhibition centre was a re-

---

1   An image of the monster's head and neck protruding from the water, taken in 1934, became the subject of much controversy. In 1994 it was revealed to have been a hoax. Nevertheless, the debate continues, as not everyone accepts the hoax story, e.g., Karl Shuker, p.87 in *In Search of Prehistoric Survivors*.

sounding success, entertaining and educating some 50,000 visitors over the summer months. The small entrance fee and revenues from merchandise sales all went toward the running cost of the LNI HQ site. The large number of visitors, and the potential commercial value, wasn't lost on some locals who could definitely see the benefits of opening a bigger centre to cash in on the monster hype. In the years that followed, Tim helped with trying to find a suitable site to create a purpose-built centre. He felt strongly the public should be shown both the facts and the massive amount of effort being expended in the pursuit of finding out the truth about the monster. Tim was quite disappointed when he realised the profits generated would be siphoned off to pockets rather than pushed back into the cost of helping with further field research. He decided to quietly withdraw himself from the process leaving others, those that held the monetary value higher than the discovery itself, to do what they do best while he continued on his own course of research.

A number of exhibition centres did eventually spring up to service the burgeoning tourist industry surrounding Loch Ness and the monster. By the early 1970s the world definitely knew about Nessie—Tim's film had seen to that. International visitors would arrive by the busload, more often than not visiting Inverness and the battlefields of Culloden first before travelling on to Loch Ness. Arriving via the easternmost end of the loch, this well trodden path led the tour buses to stop at one of the centres in the small village of Drumnadrochit overlooking Urquhart Bay. Across the water stood the dramatic castle ruins making a wonderful backdrop for tourists to snap away taking their holiday pictures for their albums. The tourists would then move on to one of the centres to learn about the loch's famous inhabitants.

Tim helped by allowing his film of Nessie to be shown at the exhibition housed in Drumnadrochit. The film became the cornerstone of the show since a moving film is far more dramatic than still photographs or static displays. The sequence went on to be played numerous times a day, and it entertained and educated thousands of visitors for many years. Rather than take a royalty for the use of the film, Tim requested the centre make a yearly donation to the Royal Scottish Society for the Prevention of Cruelty to Children.

The last of the crews left in mid September leaving Rip and Tim to sort out the Achnahannet site and get things ready for another highland winter. *Water Horse* needed to be pulled out of the loch, but before doing so Tim arranged for a few days of quiet monster watching on his own. Anchored at the Fort Augustus end of the loch he had a couple of beautiful autumn days to spend watching. With the changing of the season the sun was getting lower in the sky and would reflect brightly off the water for much of the morning. Tim knew he would need to get the sun behind him if there was to be any chance of filming Nessie. Alas, one morning while doing chores he let it slip and didn't get around to upping the anchor and moving the little boat across the bay to where he would have had a perfect view of the loch. It was a critical error; if he had repositioned he would have likely witnessed a head and neck sighting long enough to film and good enough to provide the definitive piece of evidence he so craved.

That morning Father Gregory, a monk from the abbey, had a stunning sighting of the monster. His account reads:

On Thursday morning, October 14th, I took a friend, Mr. Roger Pugh, the organist and choirmaster of St James, Spanish Place London, down to the loch side to admire the view. It was a bright sunny day, and the loch surface calm and untroubled by any boats. Standing on the stone jetty near the boat house we looked toward Borlum Bay, when suddenly there was a terrific commotion in the waters of the bay. In the midst of this disturbance we saw quite distinctly the neck of the beast standing out of the water to what we calculated later to be a height of about ten feet. It swam toward us at a slight angle, and after about twenty seconds slowly disappeared, the neck immersing at a slight angle. We were at a distance of about 300 yards, which prevented us from seeing any humps of the beast had they been exposed.

This missed opportunity floored Tim. He wrote:

after eleven years of work, through nineteen expeditions with the simple objective to spur to action, my failure to achieve it was hard to bear. The misfortune of it stunned me—it was a blow beneath the belt, which knocked me to the canvas. And yet, I had no one to blame but myself. Common sense had warned me to get the sun behind me and I had failed to take the necessary action. In all honesty I could not blame the Loch Ness "hoodoo" for that...

It was a tough one to take; years of effort and dedication and yet to miss out by such a close margin was devastating. Tim knew Father Gregory well, and had absolutely no reason to doubt him. Apart from being a man of God he was a very pleasant, sincere person Tim liked very much. Father Gregory was equally upset at Tim's near miss for he knew the work and years of honest toil Tim had ploughed through to try and unravel the monster mystery.

The day before the monk's sighting, two police offers, one an inspector the other a sergeant, reported seeing two humps moving quite fast down the middle of the loch. They filled out a sighting report form, which read:

> The first thing we noticed was a wave pattern coming towards the shore below us. The water was flat calm and a V-shaped wave pattern was coming in from about the centre of the loch. The first wave would have been about two feet high. Following the wave outwards I saw two large black coloured 'humps' about 10–12 feet behind the point where the 'V' parted. I would say there would be at least six to eight feet between the 'humps'…the impression was quite definite that they were connected below the surface. The objects were visible for two minutes at which time they appeared to go lower and lower in the water and gradually disappeared. The significant point in this was the water then returned to a flat calm condition…the objects gave the impression of two large seals or dolphins sporting, but this was only an initial impression—as time went on it became obvious that the two objects were part of one large animate object. Seen travelling over a distance of about half a mile…

The credibility of both the police officers and Father Gregory helped to soften the blow dealt by Tim's own misfortune. The two accounts, quite independent of one another, aided in Tim pulling himself up and facing the facts—once more—that it was a lottery; perhaps it's your turn to win and perhaps not. The odds may be stacked against him but he knew if he didn't even try then he was bound not to succeed. He gained hope, and knew he'd be back in 1972 to once more take up arms and battle the elusive creature.

Chapter 18

Monster Fever

Bob Rines, back in the US and encouraged by the Academy's underwater successes, put plans in place to further develop the underwater camera rig they had used during the summer. While the concept of the strobe light flashing in conjunction with a camera taking pictures was a good one, the chances of actually hitting a target were extremely slim. They needed to shorten the odds, and to that end, Bob and the Academy folks started working on a sonar-triggered camera unit. The new design would eventually consist of a sonar unit scanning the area of water directly in front of the camera rig so that when a sizable target intruded through the beam the strobe and camera would be activated. It was an intriguing plan and one that would take a whole winter's worth of development to make viable.

Bob's enthusiasm and boundless energy helped when setting up the next round of US talks. Tim's lectures in the spring of 1971 had been so successful there was a demand to follow up with another tour early in '72. This second tour would be much larger, with the British Tourist Authority helping with introductions and arranging for some TV and radio interviews. It was planned to start in New York and progress across the country to the west coast, taking in colleges and universities along the way.

Once Tim had appeared on a few radio stations, the offers started to flood in; it seemed the American public were fascinated with the subject. He talked to an audience of over a thousand students at MIT, spoke at Eastman Kodak, the Boston Aquarium, and the Scripps Institution of Oceanography in southern California. He was invited on local TV and radio in Boston, Washington DC, Rochester, Detroit, and

Chicago, and went nationwide when appearing on *To Tell The Truth* and the *Barry Farber Show*. Tim also appeared on the two biggest chat shows of the day, *The David Frost Show* and *The Tonight Show Starring Johnny Carson*. It was an impressive list and Tim enjoyed doing it. He found the Americans had a hunger for the topic and they made a good audience, asking polite, intelligent questions and enjoying the mystery of it all. And of course there was never a shortage of suggestions of how to snare Nessie, some of which were light-hearted and others deadly serious. Tim didn't mind; he took pleasure from them all.

The whistle-stop tour of the US was a resounding success in many ways. First and foremost he was able to address many thousands of people; indeed the TV audiences would take that number into the millions. He presented the growing mountain of evidence, which by then was quite compelling, to back up his own incredible film. The numerable sighting reports from credible witnesses, the scientific results of sonar showing large objects moving in hundreds of feet of water, plus the hydrophone recordings all had audiences spellbound. Secondly, Tim made a number of very interesting contacts. He met up with an old friend from Disney he'd helped out at the loch a couple of years earlier while making a children's film on the monster, and was shown around the Disney production studios at Burbank. An introduction to a very interesting man named Peter Byrne started a lifelong friendship. Peter was a former big-game hunter turned conservationist who had switched his attention to the phenomena known as bigfoot. Tim had followed the reports for a number of years of a large primate-like creature living in the Pacific Northwest of Canada and the US. He was fascinated by the 1967 Roger Patterson–Bob Gimlin[1] film showing a very large bipedal animal walking away from the camera into the surrounding forest, and naturally drew parallels with his own unusual film sequence.

The only fly in the ointment from the trip was the missed opportunity to capitalize on his appearances on two of the nation's most watched TV shows. Tim's second book, *The Leviathans,* had been updated and renamed for the American market as *Monster Hunt,* and Tim duly peddled the book on the shows. After filming *The Tonight Show,* Johnny Carson told Tim he could easily expect book sales to exceed 100,000

---

1 In 1967, Roger Patterson and Bob Gimlin filmed an ape-like creature near Bluff Creek, California. The film has never been proven to be a hoax and it continues to cause controversy over forty-five years later.

over the coming days—a very pleasant prospect, as finances were ever a concern. Sadly, the predicted sales didn't materialise, as the books had been bound with many pages back to front and had to be recalled. By the time the reprint was done and redistributed (remember there was no Amazon.com back then) the window of opportunity had slammed shut. People's memories and interest had faded over the many weeks since Tim's appearance on the shows. It was disappointing. It would have been nice to return home on a gilt-edged cloud. The trip was a reasonable financial success, even if the book sales hadn't materialised; however, such a windfall would have been a first, as Tim was still barely keeping his head above the fiscal precipice. He continued to walk the tight rope between being in the black and insolvency.

At the loch it was time for a change. The two summers Tim had spent running the show at the LNI, while immensely enjoyable, had greatly curtailed the time he was able to spend doing some monster hunting of his own. Having just spent six weeks away from his family touring the US, he didn't much like the thought of another long summer on expedition. So when Rip announced he wouldn't be returning for the season the decision to pass the burden of running the LNI onto the shoulders of others was a much easier one to make.

Rip had been Tim's right-hand man during the previous two summers, and was a tough, no-nonsense person who got the job done. One of those solid types, dependable and honest as the day is long, he could turn his hand to any number of tasks—and with an operation the size of the LNI there were many opportunities to use his plethora of skills. His contributions to the two previous summers were immeasurable: besides installing the flushing loos and hot and cold running water, Rip had been there to back up and support Tim by quietly working in the background keeping much of the equipment running and everyday operations ticking along. The two had become friends, and in the years that followed they stayed in close contact. Rip continued his connection with the loch by visiting every summer with his wife and family for the following thirty-eight years. In 1974, he produced his first Loch Ness newsletter, a collation of the year's activities, results, and sighting reports. He would send it to a small group of like-minded individuals. Rip continues to write the report to this day.

The winter months had seen Tim working with a local boat builder to design and build a replacement for the much-loved *Water Horse*.

A larger living space and working platform was needed. The boat was nearing completion and, after a successful test run on the River Thames, Tim decided to operate from the newly built boat—aptly named *Hunter*—during the coming summer.

His first port of call was to meet up with the Loch Morar Survey group; they had been running for a couple of summers and were returning for a third, and last, season. The LMS had put together some interesting findings about Morar and its ecosystem, a main point being they deduced the loch could harbour a colony of large predators. With the publication of the book *The Search for Morag*, the principal author Elizabeth Montgomery Campbell presented the survey's research and findings to date. She also listed sighting reports in a matter-of-fact way. Following the book's publication, the LMS group received a sizable donation from an anonymous benefactor to help fund the summer's research. Whether it was a direct result of the content of the book or just the fact that someone wanted to help out didn't matter; it was a very kind gesture enabling the LMS folks to spend another season investigating.

Tim always said the '72 Morar trip was possibly the most enjoyable of the entire monster hunting expeditions he ever undertook. Whether it was the enthusiastic bunch of mostly young undergrads, the wonderful and uncharacteristic tropical-like weather, or the fact that he was released from the confining shackles of running the LNI he didn't know, but he enjoyed it immensely.

The fortnight he spent aboard his new boat was also a great success. Although not quite as nimble as *Water Horse,* she made up for this one shortcoming in many other ways. Because she was an all round bigger craft, living aboard was far more comfortable and there was extra cockpit space to house the vast array of equipment she would ferry about.

For all of the good points of the Morar expedition, the lack of results marred the otherwise near-perfect two weeks. With the exception of a lone sighting of the animal's back, the LMS drew a blank on the season.

Bob Rines returned to Loch Ness as promised with an armful of tricks. His plan for the new underwater equipment had come together; some of it had been loaned to him by MIT. It was a complicated mass of electronics; sonar units, specially designed underwater cameras, and strobe lights were all attached to scaffolding and anchored in place with cables running everywhere. The new rig was positioned in about forty-five feet of water, above a steep shelf under the shadow of Urquhart Castle, at the westernmost entrance to Urquhart Bay (well away from any river outlets and salmon fishing spots). An LNI crew working from *Narwhal* helped the Academy people with the operation, and, once it was all secured in place, the rig was left to do its work.

The setup was designed for the sonar to run constantly, so when an intruder of considerable size entered the sonar beam it would trigger the strobe light to flash and the camera to take pictures of whatever had come into range. It was an ingenious piece of engineering. In the early hours of August 8, 1972, a large object moving through the sonar beam was recorded triggering the strobe light to flash and a series of pictures to be taken. Bob phoned Tim, who was at his home in Reading at the time, with the news. The film canister was kept in its original watertight container and sent back to the US for both development and examination.

The film was developed under bond by Eastman Kodak, and the sonar charts were submitted to no less than five bodies of authority for analysis. On a number of the frames, a large indistinguishable body appeared. The peat-stained water made it extremely difficult to distinguish the shape, however once the pictures had been computer enhanced (using the same process as that used to clarify images from outer space) and the peat particles removed, it revealed what appeared to be a large diamond-shaped fin, or flipper, attached to a body of considerable size.

The pictures coincided with the recording of the sonar intruder, which, for the record, showed fish as tiny dots changing to streaks when the larger object came into view—the fish were speeding up, moving away from the predator!

The image is quite incredible and, along with Tim's own film, the most important piece of hard evidence to come out of all the effort put in at the loch. The estimated dimensions of the flipper—six to eight feet in length, and two to four feet in width—fits with the sonar hit showing the intruder to be about thirty feet in overall length with signs of appendages or humps. It was a substantial creature by any standards.

The image made news headlines around the world and with Tim recognized as not only the country's, but also the world's, leading Nessie expert, the phone started to ring. It seemed everyone had woken up to the monster—again! Folks who'd not been heard of in years suddenly started to reappear: naturalists, zoologists, TV personalities, in sum a variety of people reappeared who were looking to "hang their colours"[2] on the monster and jump on the Nessie bandwagon. By now Tim had been doing this for twelve years and had seen, heard, and dealt with countless Johnny-come-lately types wanting to grab a piece of the action when the subject was once again fashionable.

The Academy's pictures stirred the pot, and with the rekindled interest came numerous offers of interviews and TV appearances. Tim's reaction was always to ask for the offer to be put it in writing, or if they were an overseas company, to come over to the loch first and take it from there. After years of dealing with the media he'd learnt to handle them with caution. Tim had a healthy relationship with a number of journalists, the ones who had proven themselves over the years to comment honestly and factually on the subject. Others, unfortunately, weren't always quite as up-front. So when unknown media companies came a-knocking, Tim took his time, wanting to know their level of understanding and commitment before accepting any offers.

About the same time that Tim received Bob's call with the news of the aquatic intruder, a Japanese TV company, Navpros, called to enquire about making a documentary based around Tim and his quest for proving the truth about the monster. A request like that wasn't new, and so, in keeping with his "come over and we'll take it from there" approach, he invited them to visit. They duly arrived in early September.

The small crew of three, consisting of a cameraman, a director and a young student interpreter, came to Reading to meet with Tim and eventually with the rest of the family. The director wanted to try a different angle on the subject compared with the multitude of documentaries already made, by using the eyes of a young child to view the mystery surrounding the monster. The director asked if I wanted to be that child. The job would entail a week at the loch interviewing adult witnesses, asking them about their experiences and what it was like to come face to face with the monster. I, of course, accepted the starring role and at the tender age of twelve sat down with my mother and what seemed like a huge camera lens just inches from my face to ask her about Dad and Nessie. I remember it all being rather strange at first as the lights

2 "Hang your colours" (or "nail your colours to the mast") is a term used when a person publicly and clearly shows personal support for something or someone.

were very bright, and with no scripts to work from I just made it up as I went along. They just pointed the camera at me and told me to ask some questions about the Loch Ness Monster.

We flew up to the loch and spent seven days touring around meeting with different folks, most of whom I already knew due to my being apart of the LNI expeditions, and interviewing them. Alex Campbell, the long time water bailiff, made a particularly interesting interviewee; he had so much experience and claimed to have seen the monster on a number of occasions over the years. We chatted away while the lights were blazing and cameras rolling in the front room of his small cottage right on the bank of loch. The old man was very gracious; he helped me with my questions and talked directly back to me in terms a child could easily understand, and any time the camera crept closer, unnerving me, he would sense my unease and help by taking over and posing a question for himself to answer.

It was a thrilling experience, hanging around with these three gentlemen from a completely different culture. The cameraman only ever wore carpet slippers, smiled a lot and chain-smoked; the director spoke in a deep guttural tone that sounded quite aggressive. The young student had to try to interpret the director's wishes and convey them to us, which he somehow managed to do via a combination of broken English, pointing, nodding, and copious amounts of smiling. It must have worked, as once back in Japan the show was apparently critically well received, gaining a number of awards. Alas, none of the Dinsdale family has ever seen it, so I still don't know if I made it past the editor's cut or whether it was my Oscar-worthy performance that won the day and perhaps the director his prize.

One has to remember at that time children didn't host TV shows in the UK; in fact adults presented all the kids' programs of the day. So when I returned to school and tried to explain what I'd been up to, none of my friends believed me. I even got my father to write a letter to the school explaining why I'd been absent—working with a Japanese TV company making a programme about the Loch Ness Monster—but not all my teachers believed it or were impressed. I didn't care what people thought, especially the teachers. I'd dealt with naysayers my whole life so it wasn't anything new to me. I'd had a great time doing something quite extraordinary, even by my family's bizarre standards, and enjoyed every amazing minute of it.

As the 1972 season drew to a conclusion, so a curtain was to close on the LNI. Sadly, after ten consecutive years of operations, the Loch Ness Investigation Bureau was to cease activities; the local authority had decided against issuing a further extension to the license for the Achnahannet site. Whether the reason was the camp had outgrown its original intended size or become too busy with all the visitors stopping at the information centre (a fact, as mentioned earlier, not missed by some locals) it didn't matter. In truth the site was too small. The narrow road had become very busy during the summer months; hordes of tourists were now driving along it, many towing caravans while straining their necks to keep an eye on both the road and the loch just in case one of its fabled inhabitants showed. With the large number of cars pulling in and out of the LNI HQ car park it was obviously only a matter of time before an accident occurred.

Other areas and potential sites were looked at and considered as a replacement, but there always seemed to be a reason why it couldn't happen or why it wouldn't be suitable. So, early in 1973, the decision was taken to finally wind up operations and dismantle the place where so many people had both visited and stayed. Over the years hundreds of individuals had given their time and energy in the quest to find Nessie, and thousands of people had stopped to spend a little of their holiday time and money to learn more about the subject. It was a moment in time when folks, young and old, from all walks of life and from all points on the globe, collectively strived to solve the riddle of Loch Ness.

The demise of the LNI was the end of an era. There were no longer any big monster hunts; the long expeditions and the collective efforts were a thing of the past.[3] Eventually, in the years to come, others took on the role of Nessie experts, positioning themselves in the limelight with the hope of gaining attention and any reflected glory that might remain from the work of others who'd gone before.

However, the closing of the LNI certainly wasn't the end of monster hunting at the loch, in fact far from it. The Academy people returned,

---

3 Although there have never been operations of the size and duration of the LNI's, other major searches have been mounted. The largest was Operation Deepscan in 1987 when twenty-four motorboats collaborated over a two-day period to trawl a sonar net over 60 percent of the length of the loch. Three unexplainable deep-water contacts were made, each said to be larger than a shark but smaller than a whale.

as did Tim and other independents, and a small group of Loch Morar Survey veterans eventually turned their attention to the Ness. This last group set up The Loch Ness Project, taking up where the LNI had left off with experiments, and eventually became the recognized port of call for witnesses wanting to report a sighting.

Tim was saddened by the loss of something he'd contributed so much to over the years. Yet, he knew that as one door closes another opens, and buoyed by the results of the Academy's efforts it was once again time to look forward and not dwell on the past.

Chapter 19

Nutter's Nook

On the home front Wendy's career with the non-profit organization was going from strength to strength. She received a new position that required working from home (quite an unusual thing to do in those days). The family dwelling was snug at the best of times, so to accommodate her new job Tim offered to build an office in the back garden. The resulting structure suited her needs well but wasn't to be her workplace for very long, as another opportunity to advance her career quickly came along and she moved out of the garden office and into a new one in the town centre—conveniently vacating the space for Tim to fill.

Finally, in 1973, after years of working from a single, small desk and sharing the children's bedroom, he could spread his wings and take the monster away from the family to create his own Nessie world. He christened the sturdy shed "Nutter's Nook." In the family's opinion it was appropriately named, as by this time we had all come to the conclusion that Tim was slightly eccentric, with a rather quirky sense of humour.

As Tim moved Nessie out of the house, so went the family's daily contact and connection with the monster. For years it had been an ever-present part of our lives, but now, with the slight physical separation, a gap began to appear. Nessie was Tim's job, and Wendy had hers; and the increasing ages of the progeny, who by now were all teenagers and by consequence striving for their own identities, widened that distance. With Tim's return to waterborne expeditions, the offers to join him at the loch weren't there, and even if they were implicit, all the Dinsdale children had their own directions and interests to pursue.

Tim spent considerable time living on *Hunter* both in 1972 and '73, continuing to assist the Academy with their underwater camera units, towing sonar and laying bait bags. He would spend days drifting, using only the power of a bow sail, or as Tim called it "snout sailing." He had cut up the old army bell tent used on the island expeditions years earlier and, with the help of a couple of bamboo poles, developed a kind of spinnaker bow sail to catch the wind and silently drift the boat along. He would spend days hugging the shoreline, working his way along one side of the loch and returning down the other. It was a peaceful way of hunting and after the hullabaloo of the last few years a welcome respite from all the media and politics that surrounded much of the LNI years.

The length of expeditions and time he would spend on the water was starting to spread out. Instead of one long trip, he'd break it up and spend a few weeks on the water then return home to do a lecture or meet an interested group before going off for another month or so of hunting. His enthusiasm for the chase never waned; the desire to prove the existence of the monster was as strong as ever, but over the years Tim had become weary of some of the more odd characters the subject seemed to attract. Operating from *Hunter,* with his own timelines, he was free to decide who to spend time with and which projects he helped out. The freedom of not being tied to one group or project in particular suited his independent nature; it also enabled him to come and go as he pleased. On the odd occasion, he did exactly that: just disappeared on *Hunter* to patrol the more peaceful parts of the loch.

The season of '73 drew interest from a variety of media groups fired up by the Academy's flipper picture of the previous year. Tim helped with another Japanese TV company making a documentary at both Loch Morar and Loch Ness, and he enjoyed the experience. A second Japanese group arrived at the loch, this one an independent company, saying they would solve the Nessie mystery once and for all. Their rather flamboyant promotions manager stated, "If your Royal Highness wish me to kill monstuh, I kill monstuh. I give the head to Queen Elizabeth, give one flipper to Princess Anne, another to Mao Tse-tung, another to…" and so on. Bold words from someone who'd never been to the loch before. The planned submarine never arrived, leaving the promoter to eat a large slice of humble pie and Nessie to keep both her flippers and head firmly attached.

The BBC asked Tim to take part in the Scottish current affairs programme *MacLeod at Large*. He also appeared on the very popular BBC's children's programme *Blue Peter,* where they had a full size model of the "flipper," as pictured by the Academy team, hanging in the studio. This gave the viewer a real feel of scale; at eight feet tall the flipper dwarfed all the presenters.

Tim published a third book. This one was titled *The Story of the Loch Ness Monster* and was written specifically for readers aged ten to thirteen years. It was a project he thoroughly enjoyed doing. He went on another lecture tour to the US, albeit considerably shorter than the previous year, staying mostly on the east coast. He had engagements in New York, Boston, Washington DC, Orlando, and Miami. In addition to the American trip, Tim spent ten very wet and cold weeks aboard *Hunter* at both Morar and Loch Ness. Another action-packed year.

Since 1968, when Tim finally gave up his position with the life insurance company to concentrate fulltime on Nessie, he'd spent 240 nights sleeping alone on board *Cizara, Water Horse,* and *Hunter.* From a truly novice sailor Tim had become a knowledgeable, seasoned boatman. He had learned to both understand and respect the moods of his environment and as a result he'd been able to stay safe and avoid any serious mishaps. The danger Loch Ness can represent was never more apparent than when Tim sent a letter home detailing a tragic accident.

On Saturday 19th May, I launched at Inchnarcardoch Bay, with the help of two pleasant young men camping in vehicles at the lay-by. One was fishing, with his own boat, and the other had an enormously powerful hydroplane, which he was testing out. While in the usual slow process of unloading gear, and preparing *Hunter* for the slip, two men arrived on the scene with a tiny wooden canoe. By now the good weather was over, and a high wind was whipping up the whitecaps. The younger man, in his mid-twenties, I should say, told me he had built the canoe himself. I warned him of the dreaded danger of capsizing into the icy water of the Ness, and the thirty-five to forty minutes to unconsciousness if unable to get one's body clear of the water; but he paid no attention. I must have had a presentiment, because I kept on about the risks, thinking he would probably tell me to mind my own business. Finally, I asked him if he had flares to let off to attract attention if he went over. He replied saying he sometimes carried them when canoeing in the sea—in short, as the Ness was only a loch he did not think it warranted them. He then pulled off in the canoe, which was scarcely visible.

About an hour later he was found floating in his life jacket, dead, and towed into Fort Augustus by a fisherman. He had capsized.

Chapter 20

Nessiteras rhombopteryx

The loch continued to surrender its many secrets, albeit slowly and with much demand on effort. Towing an enhanced version of the sonar unit pioneered by the Academy team, Tim and Bob were amazed as they watched a completely new image appear on the chart paper. Nessie was nowhere to be seen, playing her elusive game once more. However what they were witnessing was a most incredible and completely unexpected sonar response. The steep rock walls of the loch were obvious, as was the flat, plain-like bottom, but now the echo sounding was showing the walls continuing on for many hundreds of feet below the loch bed. What seemed to be happening was the new, higher intensity sonar beam was actually penetrating the waterlogged silt. The silt must have accumulated over the hundreds of thousands of years since the Great Glen Fault was created, slowly filling it. The massive cleft in the earth's crust, from mountaintop to as low as the sonar would go, was showing as 5,000 feet! The two men were astounded at what was being reproduced on the chart. Bob spent the following days checking the equipment, double-checking it, and still the results were the same. In monster-hunting terms it didn't mean all that much, but in the geological world it was a significant discovery and another Loch Ness surprise, adding still more to the mystery of the place. The discovery made Tim realise how much more there was yet to be found out about this immense and intriguing body of water.

Since the 1972 flipper picture, Bob and the Academy team continued on their underwater quest. Year on year they would return to the loch, always with a bigger and better bag of tricks to try and snare the one piece of evidence that would convince the world of Nessie's existence. Tim helped out where and when he could, offering his services and use of boats. *Hunter* was so well liked by the Academy team they eventually purchased her, leaving Tim to re-commission his much loved, but much smaller, *Water Horse*. Fuel prices had skyrocketed, so the big engine had to go as did the gas-guzzling Jag, replaced by a much more fuel efficient—and tiny in comparison—Austin A40.

In June 1975, the improved sonar system, with a combination of cameras and flashing strobe lighting, was positioned again in Urquhart Bay. Two sonar hits were made over a period of eight days, which triggered the strobe and camera units to take a series of twenty-three colour pictures. Upon development, two frames showed what appeared to be shapes of some sort, but, once again, due to the heavy peat content in the water, they were indistinguishable. After some computer enhancing, similar to the technique used on the flipper pictures, two forms could be made out. The first, recorded at the extremities of the camera and strobe lights range, appeared to be that of a strange, long-necked creature. The second, thought by some to be the animal's head, was to become known as the "gargoyle" picture due to its ugly and indistinguishable features.

Bob and the Academy team were overjoyed with the results; Tim on the other hand was a little less enthusiastic and urged Bob to be cautious with his interpretations, as the results represented not much more than cloudy grainy pictures of two objects difficult to distinguish. Unlike the flipper pictures, which were unmistakable, these two new images were, on the one hand, amazing and on the other not. A point that caused Tim concern was the camera and strobe light had been suspended under *Hunter* in mid water at a depth of about thirty-five feet. The sonar unit was a further fifty feet below, anchored firmly in place on the edge of the steep, underwater shelf. Included in the series of photographs was one of the underneath of *Hunter,* meaning at some point the camera had shifted pointing upward toward the surface. Other picture frames showed what appeared to be clouds of silt. It occurred to Tim that perhaps the boat had been blown around during the night and the free-hanging cameras could have come into contact with the steep sloping shelf and thus could have taken pictures of nothing more than decaying tree debris laying there.

Tim certainly didn't want to pour cold water on anyone's work, in fact far from it; he would go out of his way to help and support all those he felt were trying genuinely to unlock the riddle of the loch. However he knew only too well what it took to gain respect for any evidence offered of the monster's existence—and had the scars to prove it. Everything had to be fully backed up with cast iron facts. Anything less, anything sketchy, would be torn to pieces. Ultimately, a misinterpreted result—no matter how honourable the intentions—would end up doing the whole subject more harm than good and damage much of the honest work and hard-earned results which had gone before.

Tim's warnings either weren't heard or perhaps he had given them too late, because once the press got hold of the story they had a field day. The images were splashed across the front pages of the newspapers. Some of the articles gave them a scientific hearing while others did exactly what Tim had quietly predicted and tore the monster hunters, and Nessie, to shreds.

The good news to come out of it all was the flipper pictures and the summer's "head, neck, and body" and "gargoyle" pictures had created enough of a stir to waken a few from their monster slumber and helped to galvanise some new action. In the fifteen intervening years since Tim's incredible film, which had started the whole affair, a mountain of evidence had been accumulated. From film clips and still shots to sonar contact, underwater pictures, and numerous eyewitness accounts (many whose credibility was beyond reproach) it all amounted to being the right time to once again approach the scientific community and present both the work undertaken and results gained at the loch. The aim would be to get Nessie accepted as a living creature and thus, ultimately, protected. To this end, a scientific symposium was organised to take place at Edinburgh University. The following is an account written by Tim of events leading up to and following the symposium, which was added as an appendix for the fourth edition of his book, *Loch Ness Monster*.

Shortly before departure for the June series of experiments at Loch Ness, Bob Rines was invited to Slimbridge, home of the exotically beautiful 'Wildfowl Trust' where Sir Peter Scott lived. Sir Peter had been active in the early 1960s at Loch Ness, after coming to my home to view the 1960 film. Subsequently he became a founder member of the LNIB, and was largely responsible for the Linnaean Society meeting in London in 1961 at which a case for the monster was put before a panel of senior zoologists—without any noticeable effect. In the years that followed his name remained on the LNIB letterhead but he took no active part in the decade of the LNIB field expeditions as his time was occupied with work for the World Wildlife Fund and the Survival Service Commission of the International Union for Conservation of Nature, of which he had become chairman. His sudden reawakening of interest in the Loch Ness phenomenon was occasioned by the 1972 'flipper' picture, the significance of which was obvious to him; and this was heightened by the new underwater photograph obtained in June, by our Academy experiments, and also by the new long range still photography shot on 18 July by Mr. Alan Wilkins at Invermoriston Bay. Alan was a classics master at Annan Academy, and when on holiday at Loch Ness with his wife and two children early one morning he had been witness to a display of large-hump surfacing, and movement which showed them to be animate. His sequence of still photographs shot through a 640mm Novoflex lens at a range of some 3,500 yards produced measurable images, but the 16mm movie film shot through a Bolex 16mm /300mm lens was less successful.

These results were the subject of a hastily convened LNIB directors' meeting in London, and were subsequently reported in the *Field* in two splendid articles, one, "Life in the deep in Loch Ness" by David James on 23 October, which reviewed the subject as a whole, and "Monster: the four vital sightings" by Alan Wilkins on 27 November. Both were entirely clear, and objective.

Coincidently with all the happenings and news of the underwater strobe-set pictures obtained in June, Sir Peter Scott's unequivocal statements on radio that in his view the 'flipper' picture shape was distinctly like that of the plesiosaurs lent support to the visual comparison made between the photo and a diagram showing the limbs of a long-necked plesiosaur, in my April 1973 article for the *Photographic Journal of the Royal Photographic Society.* Furthermore, he painted an artist's impression of two 'Nessies,' each with a double pair of such limbs, and put this on exhibition in London, with other of his incomparable wild-fowl paintings...the stage was set.

Precisely what was to happen on that stage time was to demonstrate but, in keeping with the general atmosphere of progress and excitement, plans were laid for a presentation to be made through the

Coelacanth [Research] Committee of the Royal Society in London; but due to lead-time problems and other considerations these were modified to appear as invitations to the "Sir Peter Scott Symposium" to be held in Edinburgh on 9 and 10 December, under the auspices of the Royal Society of Edinburgh, with joint sponsorship of the University of Edinburgh and Heriot-Watt University.

The release of these 'confidential' invitations to principals of the monster hunting fraternity, senior members of the scientific community, politicians, local dignitaries from Loch Ness and others who held a particularly important view or position inevitably caused a 'leak,' which added something to the cascade of 'leaks' which had already occurred on both sides of the Atlantic, and through the printing of a small paperback for Penguin, *Loch Ness Story* by Nicholas Witchell, a law student who had put in much time at the Ness. It contained dramatic descriptions of the new underwater pictures, but was not due for publication until 11 December, and there was a strict embargo on it. Sadly, the press failed to honour this and the floodgates of unwanted publicity were opened wide which—with the whirlwind betting spree occasioned by Ladbrokes the bookmakers offering (initially) odds of 100–1: against the Monster being verified by the British Museum within one year of the bet being made—caused an immense balloon of rumour, friction and utter nonsense to inflate.

But like all artificial bubbles it could not sustain itself, and burst with a thunderous percussion, and the news that the Edinburgh Symposium had been cancelled—as indeed it had! On 1 December 1975 the Royal Society of Edinburgh issued a letter entitled:

Edinburgh Scientific Symposium on evidence for and against a 'Loch Ness Monster'

> When a Symposium to present new and review past evidence for the existence of unidentified animals in Loch Ness was first proposed by Sir Peter Scott, as Chancellor of the University of Birmingham, its local organisation was accepted by the Royal Society of Edinburgh acting in association with Edinburgh and Heriot-Watt Universities. The programme was to be that on the first day evidence should be submitted by invited experts to a restricted scientific audience and that an agreed press release would be made the following day.

It was understood that by this means, a forum would be offered for the free expression of opinions and criticisms so that a balanced assessment of the evidence could be prepared for the public.

Recent wide publicity, from prospective participants on both sides of the Atlantic, at variance with the understanding has forced the Royal Society of Edinburgh and the Associated Universities to the regretful conclusion that no useful or impartial discussion can take place at this time, and under these circumstances.

Accordingly, the Royal Society of Edinburgh, the University of Edinburgh and the Heriot-Watt University are no longer prepared to be associated with the meetings arranged for 9 and 10 December.

This somewhat haughty pronunciation was to cause an international shockwave of surprise, and a vacuum of anticlimax. The monster hunters were stunned, and not a little disgusted. So much voluntary effort had been put in by them, as usual at their own expense— but, as events were to prove, it was not the end of the matter.

David James, the unquenchable Executive Director of the Loch Ness Investigation Bureau through all its years of sweated effort has always maintained a 'press on' initiative, and as a Member of Parliament had already booked a Committee Room at the House of Commons, the Palace of Westminster, for a meeting to follow the Edinburgh affair, at which the main participants, MPs and members of the House of Lords who showed an interest would be welcomed; and he decided not to cancel it.

This meeting took place in the ancient and historic Grand Committee Room, after two days of preparations and fervent co-operative effort between the LNIB principals, the Academy team, scientists and independents such as myself who had all contributed something to the years of the search and research effort. Opinion indicated that the time had come to fight, and that the best place had been chosen.

Shortly before 8 o'clock drinks and light refreshments were served at the St Stephen's Club, at which last minute adjustments were made to the planned two-hour presentation. Acquaintanceships were renewed and the magic of human personality allowed to mix and blend together with the drop or two of elixir, known to Highlanders as the 'water of life.'

'Running order' for the presentation was listed as follows: 'Lord Craigton (in chair), 3 minutes; Norman Collins, ditto; David James, 5 minutes; Sir Peter Scott (LNIB Directors), 7 ½ minutes; Dr. Robert Rines, Dr. Edgerton, Mr. Klein, Mr. Wyckoff, Mr. Olaf-Willums, Mr. Blonder (Academy team) together 40 minutes; Sir Peter Scott (to introduce Zoologists), 5 minutes; Dr. Zug, Dr. McGown, Prof. Roy Mackal, 30 minutes; Sir Peter Scott to read extracts from the British Museum, etc., David James introduces (witnesses with supporting photographic evidence); Tim Dinsdale / film 10 minutes; R.H. Lowrie, 5 minutes; Alan Wilkins, 5 minutes; David James speaks on conservation and introduces Richard Fitter, 5 minutes; Sir Peter Scott opens meeting to questions, 35–40 minutes; Lord Craigton closes meeting, with final words by Norman Collins.

In the event, the presentation overran to some extent, leaving perhaps 20–25 minutes for questions from the floor—the press, the British Museum and other senior scientists. Members from the Commons and the House of Lords numbered perhaps a hundred, and there were many famous personalities who had at one stage or another taken part in operations, TV and radio discussions. It was a lively meeting with powerful deliveries made for both sides of the argument—feelings were aroused and spear-like verbal thrusts delivered.

There was no obvious victory for the Monster protagonists, despite the power of the technology and the facts it represented, nor was the British Museum's stand of 'not proven' acceptable to all.

The result, on the face of it, was a draw, but this did not stop the many discerning folk within the 'body of kirk' from drawing their own conclusion based on what they had seen and heard in the presentation—but as far as the outside world was concerned, still bemused and reeling from the plethora of nonsense press they had been subjected to, little or nothing had occurred, for the simple reason that the House of Commons meeting was virtually ignored by the media.

The reason for this was not immediately obvious, but there were probably three contributory factors. First, the general public was 'full up to here' with the Monster, the underwater photographs, 'Dr. Rines and his Academy', etc. Secondly, a press conference called the afternoon before the meeting, by Sir Peter Scott and Bob Rines, disclosed the Monster's new Latin name *Nessiteras Rhombopteryx* ("the Ness wonder with a diamond fin"), to be published next day with an explanatory article in the scientific journal *Nature,* along with the 1972 sonar chart, the 'flipper' photos and the startling 1975 pictures we had obtained in June some 35 feet underwater, showing what appeared to be Nessie's head, long neck and part of its great body. This meeting, in the opinion of many, distracted attention from the evening's undertakings, which were in any event late—if not actually past the 'going to bed' time for newspapers. The third factor combined with the stone-wall tactics put up by senior members of the British Museum staff convinced the press that the show was already over, and apparently not worth reporting in detail. It was a pity, because the House of Commons meeting was the most important event ever to be recorded in the Monster's long and torturous history of attempts to get the facts put forward openly and regarded seriously. In comparison with some of the rubbish that had gone before, its high technology message and equally its marvellous significance virtually went by default. However, the *Observer* on Sunday 14 December did publish a piece by Pearson Phillips, 'Nessiteras absurdum,' which must rank as a fair example of modern journalism in which the iceberg tip of facts emerge from beneath an overlay of innuendo, and blasé criticism based in the shifting sands of 'instant expertise'—but it made good entertainment and to some extent was edifying. It is as well for us at times 'to see ourselves as others see us'—but the risk with any form of entertainment writing is that wit predominates at the expense of accuracy: and of a balanced commentary. For example, (from para 2):

> It was when Dr. John Sheals of the Natural History Museum rose from a cluster of his colleagues and, quivering with professional indignation shot the whole presentation to pieces with one carefully prepared broadside. There was, he said, no evidence to show that any of the photographs taken by Dr. Robert Rines and his Academy of Applied Sciences team were of the same object let alone of a living animal. As for Sir Peter Scott's article in the scientific magazine *Nature* (awarding the beast the formal scientific name of *Nessiteras rhomboteryx*), that was spreading a 'false notion of reality' and adding disgracefully to the 'clutter of dubious names'…

This takes no account of the statements published at the presentation, from equally eminent scientists, which accept the photographs as real, and showing parts of animals. Furthermore, within the *Nature* article there are two successive photographs, taken about a minute apart, which are obviously of the same flipper-like object attached to the same body behind it.

In respect of the above, it is as well to note what these scientists actually did say about the photographs:

Dr. G.R. Zug, Curator, Division of Reptiles and Amphibians, Smithsonian Institution (a personal view): "The 1975 film includes several frames containing images of objects which possess symmetrical profiles which indicate that they are animate objects or parts thereof. I would suggest that one of the images is a portion of a body and neck and another a head...I believe these data indicate the presence of large animals in Loch Ness, but are insufficient to identify them."

Dr. C. McGowan, Associate Curator, Dept. of Vertebrate Palaeontology, Royal Ontario Museum, Canada (a personal view): "I have no reason to doubt the integrity of the investigators of the Boston Academy of Applied Science, nor the authenticity of the data...I am satisfied that there is sufficient weight of evidence to support that there is an unexplained phenomenon of considerable interest in Loch Ness; the evidence suggests the presence of large aquatic animals..."

A.W. Crompton, Professor of Biology, Harvard University, Director, Museum of Comparative Zoology: "I personally find them extremely intriguing and sufficiently suggestive of a large aquatic animal to both urge and recommend that in future, more intensive investigations similar to the type that you have pioneered in the past be undertaken in the Loch."

From these three statements alone (there are others) and disregarding all the technical evidence put forward, interpreting sonar charts, and previous surface photos, and my own film, it is clear that Dr. John Sheals' comments did not shoot 'the whole presentation to pieces,' as so glibly stated by Mr. Pearson Phillips, in whose article no one escapes ridicule for he later refers to Dr. Sheals and his four senior zoologists as "The Kensington Five," and zoologists as 'touchy people at the best of times' who when they see amateurs inventing new animals after a few summers spent dipping cameras into the water get upset.'

This latter comment may be true in the sense that none of the Academy team was trained zoologists, but in the field of technology and operations, which gained the results, we were hardly 'amateurs.' Dr. Harold Edgerton who prepared the camera strobe-set advises Cousteau on his underwater techniques, is an Honorary Fellow of the Royal Photographic Society and a Professor at MIT; Bob Rines is a physicist who trained under him, a famous patent attorney, an electronics expert and inventor; and I, as an ex-aeronautical engineer, have 21 hard-earned letters and have been a technical representative for firms like De Havilland and Rolls Royce. Our divers were both men of particular ability, with Jim Buchanan, employing his Bachelor's degree in science at Stirling University where he studied virus infections in shell-fish through an electron microscope, which was to lead to his Doctorate, and Andy Wheeler mastering the complex techniques of deep diving in the North Sea while breathing mixed gases. These facts hardly come through in the *Observer* article—but it is of little consequence.

Time alone will establish precisely what is and what is not true underwater at Loch Ness. Time and the continuing efforts of a wonderful, stimulating band of dedicated men and women—who are the true explorers, and enjoy a sense of humour which carries them through the good times and the bad.

The day the Latin name *Nessiteras rhombopteryx* was published, anagram specialists pointed out that when juggled about the letters could read 'Monster Hoax by Sir Peter S" which dismayed the Monster hunters; but Bob Rines soon came up with the antidote, another anagram—"Yes, both pix are Monsters R."

Chapter 21

A Monster's Smile

The symposium was a high water mark in gaining Nessie scientific respectability. Attended by highly qualified and respected members of the scientific community—just their presence alone indicated the importance being placed on the subject—but the heady days of the last month of 1975 were never to be repeated. As Tim mentioned, the press had had about enough of the monster; the pictures were still grainy, the public, largely fuelled by what the press fed them, was still sceptical, and the scientific community was divided as to whether the evidence was weighty enough to risk reputations by jumping onboard.

The fact remained that Tim's 1960 film and the Academy's 1972 flipper pictures were still the two strongest pieces of proof of the monster's existence and those alone just weren't enough. As Tim had always stated, short of serving Nessie on a silver platter with a sprig of parsley, clear colour ciné film was just about the only way to gain acceptance: "if you show a film of a Scotty dog running around in the back garden no one would dispute that it was indeed a Scotty dog. Equally if you show a clear colour ciné film, ideally close up, of a large aquatic animal basking on the surface waters of Loch Ness then no one "can" dispute it…"

It might have been a missed opportunity to make the monster official, but the noise it created had effects. The following summer *National Geographic* sent a team who set up an underwater station with sonar and the like. The Academy was back with more of the same sort of experiments but this time augmented with an infrared device for night-time surveying. A specially developed sonar unit was built to search the bottom of the loch with the intention of locating any bones lying in the silt. Disappointingly, no animal remains were found, but as is so often the case at Loch Ness when looking for one thing, another, totally unexpected, appears. The bottom scanning sonar experiment was no exception. Toward Dores, at the eastern end of the loch, in relatively shallow water, a series of overlapping stone circles were discovered. Whether they were a natural phenomenon or manmade in some prehistoric time couldn't be defined, but it was an intriguing discovery all the same. A second unexpected find was the remains of a Second World War Wellington bomber, which, while on a training exercise on New Year's Eve 1940, experienced engine failure and had to ditch in the loch.[1]

After the disappointment of the symposium, Tim continued on his course of independent monster hunting, spending a further eight weeks living on *Water Horse*. He continued to help out with the variety of crews and expeditions where he could and, more importantly, where he wanted to. Loch Morar gained attention from a charismatic British army explorer named Lt-Colonel John Blashford-Snell. The Colonel was a real character who had a reputation for getting things done. He was the founding member of the Scientific Exploration Society, which, since its inception in 1969, has done a vast amount of scientific and conservation work around the world. (One of the stated goals of the SES is, with each project they undertake, to leave a legacy to benefit the local community.) He had come to Loch Morar on a scouting trip to set up an army exercise for the following summer when he would be bringing a bunch of young Royal Engineers into the field. He was open-minded about both Nessie and Morag. He was intrigued to meet a local farmer and his wife who had witnessed Morag surfacing and listened, along with Tim and fellow monster hunter Adrian Shine, to their experiences. As is so often the case when meeting someone who has a monster story to tell, they came away believing the sincerity of the couple and asking the question again: Why lie? Why on earth would these gentle unassuming folks make up such a yarn? The Colonel agreed their account was indeed quite convincing.

Tim helped again with another BBC programme, and was to experience a first in his ever-growing locker of interviews by assisting with a Brazilian TV company. It seemed the monster truly had worldwide appeal. The yearly trip north to Loch Ness was preceded by a journey

1 The Royal Navy, along with the Royal Air Force Museum, eventually recovered the plane in September 1985, restored it and displayed it at the Brooklands Museum near Weybridge, Surrey, England.

south to the Cornish coast where there had been rumblings of a Nessie-like creature being spotted off the Falmouth coastline. It goes without saying such stories caught Tim's attention and, with his unquenchable appetite for a mystery, he took himself off on a quick field trip, hoping to meet and chat with some local eyewitnesses.

The reports talked of a sea creature up to forty feet in length with a long neck, small head, and the trademark Nessie-like humps being seen by offshore fishermen and land-based witnesses alike. One such witness, met by chance, was Anthony "Doc" Shiels.

The following year, Doc, travelling with his family and members of his street theatre group who were touring Scotland, happened to stop at Loch Ness. Visiting the ruins of Urquhart Castle, Doc spotted something out in the loch and took two photographs of what appear to be a reptilian-like creature's head and neck protruding from the water. The two images were quite startling, especially considering every photograph to date had either been taken at a great distance, was black and white, out of focus, or underwater, and all were controversial. Doc's pictures on the other hand were clear, in colour, and showed a lot of detail. Everything about the animal's appearance in the pictures seemed to tally with so many eyewitness descriptions: strong-looking, muscular neck tapering to a small snake-like head, eyes like slits, a mouth, and the underside lighter in colour.

Ironically, due to the brightness of the day, the water looked to be a deep, almost unnatural, blue and a strange reflection of light made the creature look as if it were smiling for the camera (this effect was later explained by a photographic expert who examined the picture and commented the smile was a consequence of light reflecting from the water onto the subject's lower jaw). The press had a field day. The fiasco of the Academy's two murky underwater pictures was a recent memory, and with the country in the grips of Queen Elizabeth's Silver Jubilee celebrations, the *Sun* newspaper seized the opportunity and ran with the picture on the front page, stating that even Nessie was making an appearance for Her Majesty.

The smiling monster and the Doc's colourful history as a professional entertainer did not make the best credentials for a plausible monster witness, and the publicity of his sighting opened the floodgates of ridicule and a torrent of disbelief. After the years of grainy out-of-focus shots, and the Academy's recent pictures, this seemed almost too good to be true. The press and public alike were more expectant of a controversial image rather than something so clear. Doc, however, was steadfast that the pictures were genuine. He claimed he had nothing to gain from fabricating pictures of Nessie and signed an affidavit to the effect that what he had seen and photographed was real and not faked in any way.

The images were certainly impressive enough, but without a point of (clear, distinguishable) reference, there were always going to be questions as to their authenticity. Beyond the affidavit and being scrutinized by film experts, there was little more the Doc could do to satisfy the questions regarding the pictures' legitimacy.

In October of the same year, Tim received a letter from Shiels which greatly helped to cement his own belief in the pictures. Tim had always been supportive, as the images struck a chord with him; he drew similarities with his own fleeting head and neck sighting from the cockpit of *Water Horse* some years earlier. On that occasion he hadn't been close enough to see details, but the shape and look of Doc's pictures were enough to sway his support and belief in their authenticity.

Tim showed his support to Doc the best way he could and that was by using one of his pictures, the more striking of the two, on the cover of the fourth edition of his book *Loch Ness Monster*.

Chapter 22

Recognition, Ridicule & Dry Dock

A great honour and a large chunk of respect was paid to Tim, in spite of his nonconformist behaviour over the previous eighteen years, when he was nominated to become a Fellow of the Royal Geographical Society (FRGS) in December 1977. His candidacy for election application read:

> An aeronautical engineer by training and, as such, fully nurtured in the discipline of the scientific tradition and scientific method, Mr. Dinsdale is a well known writer who has travelled all over the world in the connection to the investigation of life-forms still unknown to science. He is, in particular, the author of a series of very successful books relating to the leading role he has played for many years past in the investigation of the unknown creatures associated with Loch Ness and other lakes in Scotland and Ireland, as well as other countries of the world.

His election was confirmed early in the new year and Tim was, understandably, hugely flattered. This wonderful, world renowned, establishment with all its tradition, its history and famous members: Shackleton, Darwin, Livingston, to name but a few; had seen fit to recognise Tim's work as worthy of the much coveted and respected status of Fellowship. Tim was quietly proud of the honour and moreover the fact that all the work and dedication he'd ploughed into the pursuit of finding the truth surrounding the existence of the yet unidentified animals in both Loch Ness and Loch Morar hadn't gone unnoticed.

Over the years, and there had been many, Tim had become accustomed to being taken lightly or regarded as a little barmy, especially when being introduced to give a talk. He even commented in his book, *Project Water Horse,* how, on occasion, the person doing a lecture introduction would present a different attitude once in front of the audience. One moment, privately, they would be extremely interested and state their personal belief in the subject, but then promptly proceed to make a comment or two publicly that put themselves safely on the fence of indifference and put Tim—the monster expert—directly on the unsafe side with an ironic laugh and an innuendo or two. It was something he'd gotten used to and learnt not to take too seriously (after all, by his own admission he was chasing monsters) but nevertheless the "funny" comments were—whether the person making them knew it or not—slights on the serious and dedicated work that had gone on for years at the loch. The Royal Geographical Society, universally accepted as a bastion of truth and integrity, helped to put a stamp of approval on Tim's work, elevating his status from perhaps a modern day eccentric to someone whose intent and purpose was true and sincere.

The years were rattling by and with each passing one the Nessie-snaring ingenuity just seemed to grow ever more elaborate. Ideas and schemes would be tabled, and never was there a more ambitious one than the plan hatched in 1979, again by the Academy team, to use trained dolphins to track down Nessie.

A group in Florida had been experimenting with dolphins, which are highly intelligent and very trainable animals, to locate and retrieve small underwater objects. They were also being trained to carry camera units on their backs. The plan was to equip one of the dolphins with a small sonar-and-camera unit strapped to its back and send it off to search the waters of the loch for Nessie. Once the dolphin got close, the sonar would, theoretically, detect the larger animal thereby activating the lights and camera to record an underwater meeting with the monster. It was one of the more bold and ambitious plans but due to the sudden death of one of the two uniquely trained animals before even reaching Loch Ness, it was never to come to fruition.

By this time Tim had spent close to 600 days and nights on the waters of Loch Ness and Loch Morar, the vast majority of which were aboard his much loved sixteen-foot cabin cruiser *Water Horse.* And naturally, over the years he had become very much attached to the little craft (even if he did still bang his head while moving around in the tiny cabin), which had been witness to so many unique experiences and amazing discoveries. But the years of work had taken their toll on the wee boat, and, after an unfortunate incident where she was blown from her moorings crashing ashore to be holed, it was time to put her into semi retirement and invest in much needed repairs and a refit. Before her removal from monster-hunting service however there was to be one more curtain call. Prominent science fiction author Arthur C Clarke, of *2001: A Space Odyssey* fame, in conjunction with Yorkshire television, put together a thirteen-part series called *Arthur C. Clarke's Mysterious World*. The subject matter ranged from UFOs one week to the pyramids the next, missing ape-men to stone circles, amazing carved crystal skulls to questions about our solar system. All kinds of mysteries were featured with lake monsters getting their fair share of attention as well. Tim was interviewed for the series standing in the cabin hatch on *Water Horse* where he talked about the brief head and neck sighting he had in 1972. Tim went on to explain the details of his first expedition experience and subsequent filming of the monster. Watched by millions, the program aired in October of 1980 and was, on the whole, a good piece. Both Alex Campbell, the retired water bailiff, and Bob Rines contributed, among others. However Clarke's personal comments at the end of the show greatly disappointed Tim as they were presented in a way that gave the viewer the impression his opinion was in fact definitive.

> I am much more sceptical about lake monsters than sea monsters because, after all, lakes are fairly small bodies of water with plenty of eyewitnesses around them, and if these creatures come up to breathe why aren't they seen more often. And if even the Japanese can't catch them can they really exist...!

Everyone is entitled to his or her own opinion, and the questions Clarke posed about the loch being a closed body of water and small in comparison to the sea were appropriate. But, at twenty-four miles long, a mile wide, averaging 700 feet in depth—descending to over 900 feet in places—and holding over 263,000,000 cubic feet of water with a surface area of 14,000 acres, the place is certainly no puddle. Clarke also completely ignored the whole point about it being a mystery, in the truest sense of the word, and the mountain of evidence that strongly indicated there was something very strange and unknown to science inhabiting Loch Ness. Yet he boxed it into his own conventional understanding of what a lake creature should be like. He completely missed the point, preferring to revert to irony and to put himself safely on the fence of noncommittal by finishing with the innocuous, but devaluing, comment about the Japanese.

The program seemed to add to the slow decline in the level of seriousness the subject was now being given. After the steady flow of results during the sixties and early seventies, the media, and public, were primed, ready, and waiting for the definitive piece of evidence to be presented, but it never came. What did come were more blurry pictures, more shaky films, more eyewitnesses (albeit sincere), and a smiling Nessie; lots more circumstantial evidence but not the one piece, the unambiguous piece, that would have left the world without any shadow of doubt that a colony of large animals was indeed living in Loch Ness.

With *Water Horse* in dry dock (a polite way of saying stored in the garage at the family home awaiting some TLC) Tim's monster-hunting forays started to be reduced. During the off-seasons he continued to make a meagre living from lecturing about the hunt for the monster. His talks were always well received, so he decided to increase his portfolio and started to research and add subjects to his arsenal. The yeti and bigfoot were an obvious choice as he had long had an interest in the subjects, and through his friendship with Peter Byrne was connected with the up-to-date happenings. Developing the *Tungchow* piracy story by matching his own experience with others, he created "The History of the Sea" talk. And coupling the fun times he'd spent with Wing Commander Ken Wallis and his autogyro with the Goodyear airship (blimp) expedition he'd been involved with, flying over Loch Ness (memories of Operation Albatross), made a good entertaining yarn about "Airships, Sailing Ships and Autogyros." Tim added a fifth subject, returning to his aeronautical roots in the process, when he created the lecture, "Unique Episodes in Space and Aviation History."

This new array of subjects meant he could revisit many of the venues where he'd already given the monster talk. After so many years of practice, Tim knew how to both judge and engage his listeners, and he adjusted his delivery to suit the atmosphere. He particularly enjoyed talking at schools and colleges, as that's where, through experience, he could expect to be asked the most interesting and inquisitive questions. The younger audiences were keen to inquire, to know more, and were not afraid of putting up their hands to ask what they were thinking. Adults, in contrast, were often a little shy or too embarrassed to ask questions, concerned they might make themselves look foolish in the process. Nevertheless, Tim found it very enjoyable to be doing what he did well on a broader scale, and, after almost twenty-five years of talking about Nessie, found the new subjects a refreshing change.

The constant travelling, whether to Scotland and the loch or far and wide across the country delivering talks, started to take its toll. He noticed he just didn't have the energy of years past and started to book his lectures with a day's break in between. Then slowly he reduced this number still further realising he could only manage one, or at the most two, talks a week. Tim found he needed a lot more time to rest and it was annoying, as here he'd found a new revenue source. His new talks were certainly in demand, yet he couldn't fully capitalize on it. A routine visit to the doctors revealed he had high blood pressure, something he'd probably been dealing with for a number of years. The news was quite unexpected as Tim seldom smoked, rarely drank, and took exercise daily. His job for a quarter of a century involved a healthy lifestyle of living without the confines of an office schedule or the stress that tends to be associated with the corporate world. In spite of all his freedom, there was the huge, ever present, financial pressure of how to pay for his chosen way of life. Due to the sporadic nature of his income, Tim had to live with a constant overdraft—or "The Tiger Shark" as it became known—ever circling, always there, waiting to strike. As the debt increased so the shark's fin grew bigger, and got closer and closer until Tim would bundle up a whole chunk of cash and throw it at it, hoping it would be just enough to make it swim away into the distance again, for a while anyway. Tim's upbringing was one where gentlemen carried no debt; his father, even through the terrible hardships of the Internment Camp, had all his affairs meticulously in order. The years of trying to balance the books were hard on Tim's conscience; he hated being in debt and would dearly have loved the pressure of money—or the lack thereof—removed.

The news of high blood pressure signalled the end of the long, solo expeditions aboard his boat. Tim accepted it was just too dangerous to chance being alone on the water when performing below par. The loch was too unforgiving, its moods too unpredictable, and the demands on both body and mind too taxing. Yet he wasn't ready to retire from doing what he'd done for so long without reaching his goal. His aim had never changed throughout all the years of operations: to gain clear undisputable film of the monster. For this to be achieved meant he needed to be at the loch-side for at least some part of the year. So he returned to where he'd started so many summers before by spending the odd couple of weeks watching from the shore. The last purchase of a "very important piece of equipment" was a small but practical one-person caravan. It was easy to tow and even easier for one person to manoeuvre. And, as was Tim's wont of giving his equipment names, this latest addition didn't escape the honour and was christened the Perambulator Three (P3). He would trail his little home around the loch in search of a secluded nook somewhere to pull up and spend the day with his cameras at the ready. After so many years, he was a familiar face, well known by the locals, many of whom had become friends who were always willing to help in any way they could. Those with farms or a piece of land affording a view of the loch were more than happy to let him park for a few days to set up the Cyclops rig and spend the quiet of the early morning hours watching the loch. It was a return to the nomadic monster-hunting style of his first expeditions. He always enjoyed being independent, free to come and go as he pleased, to mix and spend time with whom he chose. The one downside to being based on shore again was how annoying the midges could be. After years of being on the water where he was able to escape their relentless assault, it was now back to using bug spray, keeping the doors and windows closed, and suffering the infuriating high-pitched whining late at night as one lone bug waited until Tim was almost asleep before diving in for the attack.

The 1980s were a time of change. The big expeditions had gone; the media, while still around, were generally less interested and far more in the background. And the buzz surrounding the whole Nessie hunting fraternity had shifted to a more relaxed and mellow type of pursuit. Tim greatly enjoyed developing and participating in some quite different forms of Nessie-hunting expeditions. He was thrilled to be invited aboard the sailing ship *The Eye of the Wind,* which sailed through the Caledonian Ship Canal and on to Loch Ness. This dramatic brigantine[1] vessel, had recently returned from a two-year adventure carrying over

1 Brigantine, or "Brig," is an eighteenth century term given to a sailing ship with two square-rigged masts.
2 Operation Drake, a two-year undertaking championed by Colonel John Blashford-Snell and the Scientific Exploration Society, and was the precursor to the four-year-long Operation Raleigh.

400 youngsters around the globe taking part in a variety of projects promoting youth leadership and self confidence.[2] It was a very enjoyable few days learning the sailing ship's ropes and listening to the unique sound of the wind whistling through the rigging and the moans and groans only a ship under sail makes.

He helped organize the Goodyear airship (blimp) to fly over the loch and was on board with his camera at the ready for a number of flights. The six-hour sorties were tremendous fun, although with no washroom facilities aboard one had to be extremely careful how much coffee or tea was consumed before take off! The massive blimp caused quite a stir as it travelled up and down the loch; motorists, farmers in their fields and boaters on the loch all strained their necks to get a view of the impressive sight of the enormous silver airship slowly travelling between the great cleft of mountains. Tim got some amusing shots as they flew over the ruins of Urquhart Castle with tourists looking up to take pictures of the blimp while being photographed doing so.

In late spring 1986 I had two weeks spare before flying to New Zealand for six months. Dad asked if I wanted to fill those weeks by joining him at the loch. My last visit had been with the Japanese film crew fourteen years earlier, so I agreed it was a good time to return. We stayed in a comfortable loch-side cabin at Fort Augustus, a perfect location with an unrestricted view of the loch. Tim's operating mantra of always having a camera within arm's length continued and each morning the Cyclops rig was set up and ready, just in case. I was struck by the amount of boat traffic coming through the Caledonian Canal, and as we were situated close to the point where the canal joins the loch I thought the chance of Nessie surfacing nearby was probably extremely slim.

After long, slow days of shore watching, Tim, ever mindful of the boredom that goes with hunting Nessie, decided to create some entertainment. Each evening after dinner we would float an empty Coke can away from the shore. We then took it in turns to see who could sink it using a pellet gun. Tim turned the lighthearted game into a much-anticipated evening ritual and as each day of fruitless monster hunting passed and dinner approached the previous day's loser grew eager for revenge.

Our time at the loch ended without a hint of Nessie. On the drive south I thought about the years—twenty-six by then—that Tim had been searching for the monster. I had just spent fourteen days watching the loch and while at first it was fun and the anticipation of perhaps seeing the creature exciting, my enthusiasm soon started to wane. There was no doubting shore watching was a tedious task. It was then I understood how much Tim had dedicated himself to his goal; how committed and believing he was that, given enough time, he, or another, would gain the evidence he so craved and put an end to the infernal mystery.

Little did I know it was to be the last time I would be at Loch Ness with my father; and, as things turned out, it was the last time we were to spend such special father–son time together.

Chapter 23
"I am richer for it."

In July 1987 Tim was invited to the yearly membership meeting of the International Society of Cryptozoology (ISC). It was to be a two-day symposium dedicated to "The Search for Nessie in the 1980s." To house the large gathering of world-renowned professors, biologists, zoologists and monster hunters, the Royal Scottish Museum's Natural History Department of Edinburgh opened its doors. Tim prepared a presentation on the work and results obtained in the intervening twelve years since the last such gathering at the House of Commons in London. During the proceedings, the directors sprung a surprise on Tim by announcing he'd been elected to Honorary Membership of the ISC. The unanimous agreement and warmth of the reception he received when the announcement was made was testament to the high regard he was held among his fellow adventurers. The dedication he'd shown, his integrity, and never failing sense of humour were indeed enduring qualities that had afforded him many friends around the globe. In a subsequent letter from an ISC Director Richard Greenwell, he sums up the reasons why the ISC decided to honour Tim.

> This election is in recognition of your many years of tireless efforts and fieldwork concerning the Loch Ness Monster. Regardless of whether such an animal exists or not, your dedication to the investigation, and the honesty and integrity with which you have proceeded, is unparalleled in this field. It is for this reason that the Board of Directors wishes to honour you...

Upon his return from the Edinburgh symposium, Wendy noticed

how tired and gray Tim appeared. His usual spark seemed to be missing. He talked about the ISC experience but was quite subdued, just saying how nice it was of them to recognise his contribution to the search, adding that there were many others who'd also put in much time at the loch and were as dedicated as he to finding the answer to the confounding mystery.

A few weeks later the southern counties of the England were hit by a hurricane, the first since 1703. The countryside was devastated: power lines were knocked down, trees uprooted, and roads were closed. Everything was a mess. A large branch from the oak tree that dominated our garden had crashed onto the roof of Nutter's Nook causing some damage. Tim set about cutting up and clearing away the offending branch before making the necessary repairs to his office. Again, Wendy noticed how slowly he was moving, how many times he needed to take a break, and the pallid colour of his skin. She was worried, and mentioned he should slow down a bit, suggesting perhaps he take a break from the lecturing. However, the autumn was Tim's time to make hay. The new talks were proving popular and his diary was fuller than ever before. So much for spacing the talks out! They were coming thick and fast and Tim was loath to miss any. With the popularity of the new lectures he obviously wanted to harvest as many as he could; he was well aware that once the season changed, so too would demand. After years of financial pressure he had finally found a half decent revenue source and didn't want to miss the opportunity, so continued with his busy, and tiring, lecturing schedule.

Six weeks after the hurricane, Dawn and I were off to Austria for a season of ski instructing. It was something we'd been doing for a number of years and both thoroughly enjoyed spending our winters teaching hundreds of tourists how to get down a mountainside without breaking their necks. It was early December and we were heading off to the highest parish in Austria, where the Austrian ski federation test their ski instructors, to take a ten-day ski instructors' course.

We loaded the car with everything two people would need for five months away in the mountains and said our goodbyes. I remember thinking I wish Dad would go back inside the house, as it was 5:00 a.m., cold, and he was standing in the driveway dressed in his carpet slippers and dressing gown, smiling and wishing us luck.

The small village of Obergrugl is perched on the side of a steep mountain at the end of a long, winding valley about four hours' drive from the Tyrolean capital of Innsbruck. After five days of intense skiing and evening lectures, we got a message to call home. Dawn and I looked at each other in a way only siblings can, knowing something was seriously amiss. A very cold phone booth awaited our call to England and the news that Dad had suffered a massive heart attack; the medical team didn't expect him to last much beyond the night, giving him perhaps twelve hours at the outside.

We grabbed what we thought we might need and hit the road, which was covered in both snow and ice and made for an eventful and slow journey down the valley. It was obvious that even if we drove through the night we wouldn't make it to the French ports and ferries in time, so our only option was to head for Munich and await the opening of flights at 6:30 a.m. It was a mad, emotionally charged, twelve-hour dash across Europe, and we arrived in Heathrow early the next day, only to be told that although Dad had rallied during the night, and they decided to operate, he simply hadn't been strong enough to come out of the anaesthetic.

We were simply too late. Dawn and I didn't get a chance to say our last goodbyes. It was a tough day.

We cremated Dad on a gray, wet December 21. Bob Rines cut short his business trip to Beijing to attend the funeral and gave the eulogy. People, most of whom we all knew, came back to the family home. There was the odd unfamiliar face, and while chatting to one he explained he'd been at de Havilland's Digswell House with Dad directly after the war. Dad had, on occasions, talked to us about "the boys," the group he would go to dances with and have a merry old bachelor time of things. The friend said he'd followed Dad's monster career with interest, stating that he and others who knew Tim from those days greatly envied his courage to follow his own path. Many of his peers had gone on to achieve high positions in industry, yet they all—to a man—admired my father's life choice.

Tim's passing was naturally a great shock, as is a sudden death to any family. We reeled in the finality of it, fighting the thoughts that he was just away on another one of his trips and would be home in time for Christmas. Wendy found a pile of Christmas cards he'd written and decided to send them, including within them a note telling folks of Tim's untimely death. His own message to all his friends around the world was one of optimism and anticipation for a successful hunt in the coming summer months.

Tim's enthusiasm for the chase had never waned. To the very end he maintained his belief that he, or some other person, would ultimately capture the elusive animal on a clear sequence of ciné film, thereby finally putting an end to the riddle. The years of effort, of endless uncomfortable cold nights and wet miserable days never dampened his spirits. The counter balance of stunning scenery, and living a life filled with adventure and passion was more than equal to any personal discomfort. Of course over the years he got despondent at the lack of results, frustrated with the endless hours of toil with little to show for it, and the stonewall tactic of the scientific establishment. But through all the thousands of hours spent on his lonely quest Tim was happy, content at the choice he'd made years earlier to step out of the mainstream and into a world which proved to be anything but predictable. The price he paid for living a life of adventure was, in his mind, small in comparison to the payback he received. To quote the opening lines of his book *Project Water Horse:*

> It was the twenty-seventh day of September 1967, and I was conscious of becoming technically a year older. At forty-two, with a family of splendid youngsters to support, my adult behaviour during the past decade left much to be desired, when judged by the standards of convention. And yet despite that I was happy. Harassed but happy, penniless but free, and treated with good humour by the progeny, because I had given up a career in aeronautics to embark on a form of madness of which they thoroughly approved—the pursuit of a legendary monster…And here I was 'at it' again…

Epilogue

The months immediately following Tim's passing Wendy continued to receive lots of well-meaning letters of support. On more than one occasion writers mentioned their personal regret that Tim hadn't managed to improve on the incredible luck of his first expedition and gain that definitive piece of evidence, thus confirming the animal's existence. It was indeed a tremendous disappointment and heart wrenching for the whole Dinsdale family that Tim died before he'd succeeded in his own goal of capturing unequivocal proof of Nessie. After the years of toil, of sacrifices, of time spent away from his family, it all seemed rather unjust the story hadn't cumulated with a classic Hollywood-style ending.

It has been suggested that perhaps Tim *had* in fact already achieved his goal, as the film he took in 1960, on his very first expedition, is quite simply astounding. Nothing like it had ever been seen before, or indeed, has been bettered since. All the hard scientific facts surrounding the sixty-second sequence imply, by process of elimination, whatever was moving through the water on that April morning was in fact an animate object of considerable size swimming at a speed greater than any creature known to inhabit the loch. A little later on that same fateful day Tim sat again in the exact spot and filmed a second sequence. This time he focused his camera on a fourteen-foot wooden fishing boat crossing the loch, following a similar track to the one the object had taken only an hour or so earlier. The comparison leaves the viewer in no doubt the boat is exactly that, a boat, pure and simple; however the other, the black triangular-shaped hump, is anything but.

Perhaps Tim's film is the best Nessie evidence we're ever going to get. I know he'd be heartedly disappointed if that turned out to be so. Tim always maintained what he saw through his binoculars that morning is represented rather poorly on the film. At extreme range the object on camera is little more than a black cone zigzagging across the loch, while what he saw with his eyes aided by magnification was

> a long oval shape, a distinct mahogany colour. On the left flank a huge dark blotch could be seen, like the dapple of a cow. For some reason it reminded me of the back of an African Buffalo—it had fullness and girth and although I could see it from end to end there was no visible sign of a dorsal fin. And then, abruptly, it started to move. I saw ripples break away from the further end, and I knew at once I was looking at the extraordinary humped back of some huge animal ...as it swam across the loch it changed course, leaving a glassy zigzag wake. And then it started to slowly submerge. At a point two or three hundred yards from the opposite shore, fully submerged, it turned abruptly left and proceeded parallel to it, throwing up a long V wash. It looked exactly like the tip of a submarine conning tower, just parting the surface, and as it proceeded westwards, I watched successive rhythmic bursts of foam break the surface—paddle strokes, with such a regular beat I instinctively started to count: one, two, three, four, pure white blobs of froth contrasting starkly against the black water surrounding, visible at 1800 yards or so with the naked eye..."

At the very start of this book I asked the question why a well educated family man with a solid career in a cutting-edge industry would turn his back on the assurance offered by employment and risk his family's financial security to go and chase a mythical monster in the highlands of Scotland. I wanted to know what the initial drive was, what was the spark that ignited Tim's sense of adventure to leave his pregnant wife with three small children and go off for a week to hunt a creature which, at the time, would have been akin to looking for a unicorn.

The answer as to why he ventured to the loch in the first place is rather straightforward, and nothing more exciting than an inquisitive mind wanting to know more—and perhaps a healthy sense of adventure. However the sequence of events of filming the creature and presenting his findings to the world via the BBC's *Panorama* program, were almost custom designed to snare Tim's curiosity.

Naturally—or naively—he anticipated the scientific establishment would be just as awe-struck as he and many others were about the film. But his evidence was met with apathy. The reaction of some of the country's most prominent zoologists of the day he found a great disappointment. There seemed to be a blanket refusal to even consider any possibility of a colony of large unknown animals living in the water of Loch Ness; indeed they flatly refused to even acknowledge Tim's film as evidence.

This stubbornness, their indifference, and in fact the arrogance of the scientific establishment dismissing the subject outright even before a case had been put forward, annoyed Tim. He knew what he had seen and filmed wasn't anything that could easily be explained; it wasn't a boat—that was obvious. And it wasn't a submarine, so what was it? The only logical explanation was an animal of some unknown description. And yet here the very people Tim had turned to for their learned and expert opinions were spurning his evidence with such ease it was an attitude boarding on disdain.

This rebuff just added to Tim's dogged resolve of bettering his already incredible film. The success of his first trip to the loch had left Tim in absolutely no doubt there was an extraordinary—new to science—creature of quite stunning proportions living in the waters of Loch Ness. After all, he'd seen it! And so when it came time to pack up the car and head north again it was always with a genuine feeling of optimism, of perhaps this trip, this expedition would be the one when the dark foreboding waters of the loch would finally reveal its centuries-old secret.

It also has been suggested Tim became obsessed with the monster. I don't believe that to be the case, or no more so than perhaps a committed athlete would when striving for a goal or even a business person trying to clinch the next contract. Tim's metamorphosis from a respected career professional to full-time monster hunter happened over a period of eight years; after numerous expeditions he slowly became competent at his craft. Tim literally became a professional monster hunter, unusual as that may sound, but chasing Nessie eventually became his fulltime, serious, occupation. Tim recognised what he was doing was probably one of the "most extraordinary [careers] on record, but in the human sense, it is perhaps one of the most rewarding," and would readily admit his occupational choice was anything but conventional. He was one of the first to poke fun at himself, but not at the subject or the serious on-going work at the loch. There were a few honest, steadfast and committed individuals dedicated to the search who, along with Tim, expended a huge amount of time, energy, and money, all of whom had his utmost respect.

Tim once said he thought the term "monster" was a rather dramatic and perhaps unfortunate word choice to describe the animal, as a monster seems to conjure up an image of a mythical beast sending one's imagination running to the pages of Arthur Conan Doyle's *The Lost World*. He felt if a more scientific title had been attached, Nessie might possibly have been afforded a fairer hearing from the zoological fraternity.

But the term has stuck and done nothing to harm the thriving tourist industry that has blossomed around the loch—and its monster. The Nessie legend has grown to become a worldwide phenomenon attracting over 500,000 visitors to the area annually, many wanting to know more about the infamous inhabitants, their interest ignited by years of perplexing evidence that just won't let the subject die.

Prior to 1960 there had been occasional pictures appearing in the press claiming to be Nessie—some of which have since been revealed as fakes while others continue to confound—but Tim's film can be traced as the catalyst for the modern era of Nessie hunting. The LNIB was formed in 1962, following Tim's appearance on the BBC's *Panorama* program. That same year Birmingham University scored the first sonar "hit," and they repeated their success in 1968. The following year the Pisces submarine made mid-water contact with a forty-foot "something," which moved off when approached; and Bob Love's expedition claimed a sonar contact in the same summer watching the "target" swim around for over two minutes.

In 1970, Marty Klein, trawling his side scan "tow fish" sonar unit from the boat *Water Horse,* picked up a large object swimming in deep water just off an area known as the Horseshoe. And in 1972, The Academy of Applied Science recorded a thirty-foot "intruder" as it passed through their sonar beam in Urquhart Bay, when they captured what has become known as the flipper picture. The team repeated their underwater success in 1975 with two more intriguing pictures.

The LNIB had their own share of success with the occasional photographic sequences being taken throughout the years of the organisation's existence; probably the best of these was shot in 1967 by

Dick Raynor showing a large wake moving at speed, keeping pace with the pleasure boat *Scott II*. And in 1977 a "smiling monster," the head and neck colour picture taken by Doc Shiels.

These form just a small portion of the body of Nessie evidence that exists. There are many hundreds, perhaps more than a thousand, eyewitness accounts claiming to have seen the animal stretching back over decades and indeed centuries.

It is debatable whether or not there would have been as much interest in Nessie if Tim hadn't taken what has become an iconic piece of film. It was a chance reading of an article that changed his life. That moment in March 1959 lead to twenty-seven years of search and research. There was laughter, mystery, and adventure in the form of fifty-six monster-hunting expeditions. His desire was to ultimately capture his quarry on clear, close-up, colour ciné film for no other reason than proving the "truth for truth's sake."

In his pursuit, and in the eyes of modern-day society, he sacrificed much: career, financial security, a larger home, private education for his children. But these perceived losses were never felt by any of his family. There was never a feeling of going without; in fact the opposite was true. The variety and richness of experience living with Nessie brought the Dinsdale family was truly unique. Summers spent by the loch side, meeting all sorts of wonderful characters, participating in exciting experiments marvelling at the occasional results, laughing at the sillier moments, yet always with an anticipation that today, tomorrow, or next week we might finally succeed and capture the mysterious beastie on film; and *then* serve her up to the sceptical scientific community—sprig of parsley and all!

It goes without saying we had to endure a considerable amount of ribbing from school friends and on occasion had to explain Nessie to the less worldly. And yet, in spite of the occasional naysayer our world was one full of adventure. Although my family's association with the loch has diminished over the years and our connection to the mystery is at considerable distance nowadays, Nessie is always there in the background, and via Tim's film we are, and always will be, indelibly linked to the subject.

In life, Tim counted himself a lucky man. He survived the Second World War by virtue of the conflict ending before his pilot training was complete. He had good fortune in his early professional life, which happened to coincide with new and exciting, often ground breaking, developments in aeronautics. His second, quite unusual profession was self-made; he became the trailblazer for both the volunteer and professional monster hunter at Loch Ness. It was an occupation he thoroughly enjoyed and was well aware of the richness of the life he was living, a wonderful tapestry of experiences, colour and variety of people all trying to achieve something quite extraordinary.

And for all the hullaballoo that surrounded monster hunting and Tim's status as a world-renowned Nessie expert he was still just our dad. With Wendy's time taken up doing her job, Dad was the parent we'd come home to during the school term. For the summer months, when Tim was away on expedition, Wendy had an arrangement with her employer letting her take all her annual leave in one go and to also add on a month's unpaid vacation time. Combining the two periods allowed her to look after the family during the long summer recess, enabling Tim to go off and chase the monster. It meant that growing up we kids had a wonderful variety in parenting style; mother steadfast, career minded and conventionally responsible, yet far less strict than Tim. Father, on the other hand, while insisting on good manners, politeness, and respect, wasn't the best example when it came to toeing the line of conventionality. He would marvel at his own freedom, having broken the corporate shackles years before, and chuckled at the thought of being stuck in a stuffy meeting room full of a thick blue haze (it was commonplace for both cigarettes and cigars to be smoked during business meetings back in the 1950s, 60s and 70s) when compared to the exhilaration of being on the loch in *Water Horse*. As my siblings and I grew and developed into adults we just accepted Dad's unusual profession, and were always quietly proud of his resolve and integrity. He was equally—and often more so—proud of his offspring and their chosen paths in life: Simon in the military, followed by a thirty-year career in the police force; Alex, a military nurse; Dawn, a teacher; and myself… well I've always been the odd one out when it came to career. Like my father I tend to follow my heart, which has led me into some wonderful life adventures, and as such I haven't been much of a conformist—but when you look at my role model, how could I expect to be!

Tim loved his family without parallel and recognized if it weren't for Wendy's unwavering support none of his adventurous lifestyle would have been possible. His family loved him back in equal measure and accepted we were just a very ordinary family with a father who did something quite *extra-ordinary*.

The cost has been great, at a private level seemingly impossible to meet in time and money, and yet, in meeting it, by some strange alchemy I am richer for it, and my family no less independent. And so too, the Monster justifies itself in terms of opposites; because I do not believe it is in itself important. Dramatic, extraordinary, exciting, a zoological wonder perhaps, but not important, in the sense it is only an animal—like an elephant or for that matter a cow, which is equally as marvellous.

But in the way it relates to our scientific society, it is of enormous importance. In the case of embarrassing unexplained phenomena, science just 'doesn't want to know'—and for this reason it is imperative that voluntary work continues at Loch Ness. We stand on new frontiers of discovery, which will test the credulity and courage of man, and his ability to adapt will depend on his mental flexibility. We must have this type of mental outlook, and at Loch Ness we have such a rare opportunity to demonstrate the need for it…

—T.K. DINSDALE 1924–1987

A Gallery of Monster Hunters

It struck me, as I wrote these pages, how many years have passed since the monster-hunting heydays of the 1960s and 70s. The names of people I remembered so well from my childhood kept popping up; yet these folks, while so familiar to me, would mean little or nothing to an interested reader now a generation removed from the loch's wonder years of investigation.

During that twenty-year span when the Ness came under so much scrutiny, the level of interest in its secrets stretched around the globe. The subject, and perhaps the chance of becoming forever known as the person who proved the creature's existence, lured many—well educated, often expert and occasionally world leaders in their fields of research—to use their knowledge and wits to unravel the mystery that is Loch Ness.

To enlighten the reader somewhat, I include here what one might call a rogues' gallery of monster hunters from times past. As I suggested at the start of this book, there were many interesting and fun characters that played some part in the search and to name them all would be an arduous task indeed. So I've picked a few names, some you've already read about while others will be new, but, nevertheless, I believe of interest.

Loch Ness Investigation Bureau (LNIB)

• David James MP, MBE, DSC

Educated at Eton and Oxford before joining the Royal Navy, David James commanded a gunboat during the Second World War and was taken prisoner by the Germans. He quickly escaped only to be recaptured, but then attempted a daring second escape. The stakes were high, because if he were caught again he would have faced certain death. He succeeded, eventually making his way to Sweden. The account of his brave escapade is detailed in his excellent book, *Escaper's Progress*.

David went on to become a prolific author and occasional filmmaker. In 1959 he became a Member of Parliament. He was the leading force in the development and then creation of the Loch Ness Phenomena Investigation Bureau. During the years of the LNIB operation, 1962–1972, he was the bureau's political face, the person banging the drum to get support and donations to keep the whole thing going. At the general election in 1964 he lost his parliamentary seat by a tiny margin of just seven votes; it was rumored his involvement with Loch Ness and its monster had hindered his re-election. David eventually won a seat in the 1970 election, which he held until he retired from politics in 1979. He passed away in 1986 at just sixty-six years of age.

• Sir Peter Scott CH, CBE, DSC and bar, MID, FGS, FZS, Olympic medalist

Sir Peter Scott was the son of legendary naval officer and Antarctic explorer Robert F. Scott. A keen naturalist, Sir Peter founded the Severn Wildfowl Trust in 1948, and was also a co-founder of the World Wildlife Fund. He became a trailblazer for wildlife conservation, receiving awards for his work. One of his many contributions was helping to place a restriction on whaling in the waters surrounding Antarctica. He led the way in bringing nature to the TV screen, and was also a gifted artist whose wildlife paintings are much coveted.

Sir Peter gained his initial interest in Nessie during a private showing of Tim's film only a few weeks after it had been taken, and he subsequently became a co-founder of the Loch Ness Phenomena Investigation Bureau. He famously captured his vision of Nessie in oils shortly before giving the creature a Latin name, *Nessiteras rhombopteryx*, the Ness wonder with a diamond fin. Sir Peter died in 1989 at age seventy-nine.

• Ken Wallis MBE, DEng (hc), CEng, FRAeS, FSETP, PhD (hc), RAF (Retired)

At ninety-six years of age (at time of publication), Ken is the oldest aviator still flying in Great Britain and still flying his self-designed, homemade autogyros. He has held more than thirty world records flying his versatile little machine, some of which he still holds. Ken was a Second World War bomber pilot who gained a reputation for being adept at bringing his plane home no matter what the condition. A true inventor, he has a long list of design patents to his name; but one that slipped through his hands was his creation of a slot car racing track. During the war, for some fun with his friends, Ken built a toy race track using his own designed three-inch model cars. Some years later he asked a friend to register the patent while he was away overseas. Alas, the friend forgot and the following year, 1957, Scalextric appeared. Ken's invention is still in perfect working order today, and he maintains his design is better than Scalextric's, as his cars have turning front wheels.

An enthusiastic member of the LNIB team, in 1970 Ken made eighteen flights and clocked nineteen hours fifty minutes flying time over the loch. He said when he looked down at the loch it was like looking into an inkwell, the water was that murky. Ken recalls a warning issued during World War Two from a naval pilot based at RAF Lossiemouth in Scotland, saying no one should go for a swim in Loch Ness as a crocodile was living there!

• Dick Raynor

During Raynor's first stint with the LNIB in 1967 he shot a piece of film showing a fast-moving wake keeping pace with the pleasure boat *Scott II*. From then on Dick was a regular summer volunteer with LNIB and helped run the Achnahannet headquarters the final year of operations in 1972. He is still to be found around the loch driving tour boats while continuing his own investigations.

American Ingenuity

• Harold "Doc" Edgerton, Professor of Electrical Engineering, MIT

Doc's achievements and pioneering work in both strobe light photography (which allowed rapid events to be captured on film) and sonar development are great indeed. He worked with world-renowned underwater explorer Jacques Cousteau custom designing undersea photography equipment, and built a first of its kind side scan sonar system which helped to locate the American Civil War battle ship USS *Monitor*.

Edgerton lent his extensive knowledge of underwater photography and considerable skills with sonar to the Academy of Applied Science. Together they teamed up and built the camera/strobe/sonar rig that took both the 1972 and '75 underwater pictures in Urquhart Bay at Loch Ness. Doc passed away in 1990 at eighty-six years of age.

- ## Dr. Martin Klein PhD, MIT Class of '62

Tutored by Doc Edgerton, Marty demonstrated his brilliance in electronic design in developing side scan sonar. The US Navy became a customer, at one point using his technology to locate a lost nuclear submarine. The list of finds accredited to Marty's "towfish" technology is quite impressive, including two 1812 warships in Lake Ontario, the Wellington bomber in Loch Ness, and of course the most famous of them all, the *Titanic*.

- ## Dr. Robert "Bob" Rines

Founder of the Academy of Applied Science, Bob's list of both professional and personal achievements is quite remarkable. An MIT grad, he also received a law degree from Georgetown University and went on to earn a doctorate from Chiao Tung University in Taiwan in 1972. Bob was a well known intellectual property lawyer, composer, accomplished musician, and inventor with a large number of patents to his name. His first visit to the loch was in 1970; he returned many times, making his last appearance in 2008 a few months before he passed away at age eighty-seven.

- ## Ike Blonder, Co-founder of Blonder-Tongue Laboratories

Ike, along with business partner and friend, Ben Tongue, was a pioneer in the field of TV, cameras and equipment, registering thirty-nine patents. Ahead of his time, Blonder was a leader in 3D TV technology, in 1989 launching an experimental 3D channel, and was a leading figure in the development of pay-per-view for cable TV, among many other innovations in the entertainment industry.

Ike was a fixture with the Academy of Applied Science team, visiting the loch numerous times during the 1970s and 80s. He was another electronics and image expert who contributed much to their efforts. Ike was ninety-two when he passed away in 2008.

The Independent Nessie Hunters

- ## Dr. Henry H. Bauer, Dean (retired), Virginia Polytechnic Institute and State University

Dr. Bauer, emeritus professor of chemistry, has a long history of researching pseudoscience topics, and became quite a controversial figure. His list of publications includes some subjects not always readily accepted within the scientific community. A firm supporter of Tim's Nessie film, Henry was an occasional visitor to the loch, and wrote a book on the subject of the monster.

As a founding member of the Society for Scientific Exploration (SSE), Henry was instrumental in setting up the Dinsdale Prize (www.scientificexploration.org/dinsdale.html), to recognize "significant contributions to the expansion of human understanding through the study of unexplained phenomena."

- ## Dr. Roy P. Mackal, Biologist (retired), University of Chicago

Dr. Mackal, with a background in biochemistry, virology and zoology, was a familiar face at the loch from the mid 1960s onward. He was a founding member and vice president of the International Society of Cryptozoology. His interest in unexplained creatures wasn't reserved solely for Nessie; he had a healthy curiosity in the many reports of strange animals from all over the world. The story of a dinosaur-like creature coming out of the Congo in central Africa attracted Mackal's attention and in the early 1980s he mounted two expeditions to the country's remote northeastern region. He recounts his African adventures in his book, *A Living Dinosaur? In Search for Mokele-Mbembe*, and in 1976 also wrote a book about Loch Ness, titled *The Monsters of Loch Ness*.

- ## Ivor "the diver" Newby

An accomplished sports diver, Ivor was one of a dedicated band of independent monster hunters who, along with his boat, *Kelpi*, was around to lend a helping hand where needed, although he was not particularly affiliated with any one group. Coincidently, Ivor made his first visit to dive in the loch the same week Tim filmed the monster in 1960, yet continues to profess it wasn't he that Tim filmed.

- **Nick Witchell**

Nick, a quiet regular at the loch during his younger years, read law at Leeds University. After a number of summers observing the various teams working at the loch, he published an excellent book on the subject, *The Loch Ness Story*. Nick went on to become a highly respected journalist, reporting from areas such as Northern Ireland during the troubles there and the first Gulf War. A longtime news anchor for the BBC, Nick is the corporation's royal correspondent.

- **Peter Byrne, F.R.G.S.**

Big game hunter, conservationist, Himalayan explorer, yeti and bigfoot expert, are all honest descriptions of this regular Indiana Jones style character. In the late 1940s Peter worked in northern India on tea plantations. While there, he came into contact with locals telling yarns of a large, hair-covered bipedal animal living in the high Himalayas. Working with a wealthy American, Peter lead a three-year expedition into remote regions of the country looking for signs of the fabled beast. Footprint sightings and a strange mummified hand found in a monastery were the sum total of their findings; not sufficient to prove the animals' existence but just enough tantalising evidence to convince him there was something extraordinary to be discovered. Sound familiar?

Peter initially made a living working with tourists hunting tigers in Nepal, but after years of trophy hunting he changed his focus to conservation. He was instrumental in setting up the International Wildlife Conservation Society, turning the vast area of Nepal he would once hunt into a game reserve. The Pacific Northwest was Peter's next port of call. He had heard reports of a large primate said to be living there and so went to investigate the mystery of bigfoot, subsequently leading two long-term projects to locate the creatures.

Peter visited Loch Ness in the early 1970s, helping Tim and others with experiments and general monster hunting activities, becoming part of the renegade band of independent Nessie hunters.

Peter Byrne is a Fellow of The Royal Geographical Society, London, England; Member Emeritus, The Explorers Club, New York, NY; Member, The Academy of Applied Science, Boston, MA; Honorary Member, The East African Professional Hunters Association; and Honorary Executive Director, The International Wildlife Conservation Society, Inc.

Other

• Ken Peterson

As one of the original Disney animation artists, Ken worked on *Snow White and the Seven Dwarfs* in 1937 and went on to have a long and distinguished career at the famous studios, working on the production of some animation classics, such as, *Alice in Wonderland, Sleeping Beauty, Lady and the Tramp*, and *101 Dalmatians.*

Ken arrived at Loch Ness in the summer of 1969 with his crew and shot what is termed a "short." What resulted was a twenty-four-minute educational style documentary with an animated Nessie doing much of the talking. *Man, Monsters and Mysteries* is an extra feature on the DVD of another Disney classic, *Peter's Dragon*, and also can be viewed on YouTube, <www.youtube.com/watch?v=Q72eI_EA4rc>. In the film, Tim talks about his conclusions regarding the animals' makeup, and there's a clip from his film and another of Tim looking menacing with his crossbow. As well, many characters mentioned in this book are featured, including: Ivor explaining diving in the loch; David James addressing a new bunch of volunteers at the LNI Achnahannet base camp; Ken giving a wave as he takes off in his autogyro; Roy Mackal talking about taking a skin sample; Dan Taylor explaining a strange occurrence from the conning tower of his yellow submarine; and long-time water bailiff, Alex Campbell, recounting his story of seeing the monster in 1934. Ken passed away in 2000 at age ninety-one.

Bibliography

Dinsdale, Tim. 1966; revised 1976. *The Leviathans*. Routledge & Kegan Paul.

————. 1972. *Monster Hunt*. USA edition of *The Leviathans*. Acropolis Books.

————. 1973. *The Story of the Loch Ness Monster*. Target, a division of Universal-Tandem Publishing.

————. 1975. *Project Water Horse*. Routledge & Kegan Paul.

————. 1982. *Loch Ness Monster*, fourth ed. Routledge & Kegan Paul.

Index

Acknowledgements

This is a story I've long wanted to tell, yet had neither the time nor self-belief to do so. The opportunity to tackle this project came, as is so often the case, out of adversity. Sudden unemployment presented me with the time, a redundancy payment supplied the financial cushion, and encouragement came in the form of a dear friend, Helen Gould, whose enthusiasm for the subject was infectious. While present at her family's 2009 Thanksgiving dinner I pronounced that indeed my quivering quill was about to hit paper. The very next morning I sat looking at a blank computer screen and stared, and then stared some more, as where to start wasn't at all clear. "At the beginning, silly," came the shout from Helen—and so it was.

I would like to thank my darling sister Dawn who is the custodian of my father's materials and as such has been on hand to sift through the mountains of papers, checking dates and historical facts to make sure I got things right. My brother Simon, a humorous raconteur himself, was always available to add forgotten snippets to a tale or situation, and along with Dawn put plenty of hours into collecting the story-enhancing mound of photographs. Alex, my oldest sister, was also greatly encouraging and available with her own memories of those times we spent on expedition.

I would like to mention and thank my Aunty Felicity. Without her memories, some quite painful to recall, much of my father's early life in China—and the Tungchow incident—would have been so much less to tell.

And to those who have kindly agreed to allow their photographs to be reproduced here in these pages, thank you. For the illustration of the loch and places of interest pertaining to our story, credit has to be paid to the artistic skills of multi-talented Ricki Baxter.

The publishing of this book would not have happened without the initial editing skills of Joanna Gould (yes, that's Helen's sister) and then the dedication of the entire team at Hancock House Publishers, especially Theresa Laviolette and Ingrid Luters who together helped to make this a better book. David Hancock needs to be mentioned and thanked, not only for publishing this book but also for keeping a channel open for unconventional subjects to make it to print.

My final thank you is saved for my mother, Wendy, who has graciously allowed me to talk about our family so openly and regurgitate so much history. I am also grateful for her permission to incorporate passages from my father's books, as this allows the reader to "hear" my father's words and as such build his character in his or her mind's eye.

This project has been a labour of love. Not many sons get the opportunity to write about their dad's life; I have enjoyed every moment of doing it. Thank you to one and all for your help, and more importantly, belief.

About the Author

Angus Dinsdale has monster hunting in his blood. Born the very day his father presented his famous film of Nessie to the British nation on the BBC program, *Panorama,* Angus grew up in a household where searching for the Loch Ness monster was his father's full-time— and perfectly acceptable—occupation. Subsequently Angus's childhood was full of adventure as he took part in many monster-hunting expeditions to Scotland. He developed a love of the outdoors, sport and travel, interests that have taken him all over the world, turning his two key passions of skiing and rowing into adventures all of their own. He now lives in Vancouver, Canada where he has found the inspiration to write, surrounded as he is by the essential playground of an adventurous life: majestic mountains, elemental wilderness and stunning lakes—which might even have a Canadian Nessie of their own…

Tim Dinsdale's filming of a Loch Ness "monster" in 1960 set off more than two decades of intense searching and enormous publicity; this book recaptures the excitement of those years and reminds us of the wealth of solid evidence that Nessies are real creatures.

Through three decades of expeditions, Dinsdale kept a clear-headed, critical appreciation of that evidence despite the staggering frustration of not bettering the 1960 film. Tim's integrity and strength of character stand in stark contrast to the frustrated others who try to explain away the evidence as sturgeons, birds, otter-boards, etc., and who, saddest of all, claim that the 1960 film shows only a boat. Tim's film is now available at this book's website, allowing everyone to see what nonsense this "boat" explanation is.

Tim was always clear about the importance of his quest and he took it with appropriate seriousness, but he never took himself seriously. He was a delightful companion, full of good humor, a wonderful exemplar of free spirit coupled to conscientious responsibility.

Tim's dedication and integrity influenced innumerable lives. His work stimulated a career change for me, very much for the better; and I find myself still grappling with the conundrum that floored Tim 50 years ago: Why does science ignore so much that's of such great interest?

DR. HENRY H. BAUER
Emeritus professor of chemistry and science studies
Virginia Polytechnic Institute and State University

Tim Dinsdale was one-of-a-kind. A special man from a special generation. He combined a sense of adventure, incubated by his early travels and hardened by his wartime service, with an infectious curiosity about life and a bloody-minded determination to puncture establishment pomposity. In 1960 he found a focus for these instincts – and his long and determined quest at Loch Ness was launched.

I met Tim for the first time in 1970. That was the summer when he was running the Loch Ness Investigation Bureau site at Achnahannet. It was my first visit to the loch. I'd become intrigued by the mystery. Looking back, I realise these were the first stirrings of that journalistic instinct which was subsequently to take me to a working life with BBC News.

Loch Ness was such a good story. It had everything: romance, mystery, adventure and the prospect of a startling discovery. Tim was one of a band of wartime veterans who were drawn to it. I suspect they may have found civilian life a little dull. Loch Ness was a marvellous diversion – except that, for Tim, it became so much more.

He was a delightful man whose life took an extraordinary turn. It was a pleasure to have known him, his wife Wendy and his family, and to have been able to re-live his great adventure through the pages of Angus Dinsdale's absorbing biography.

NICHOLAS WITCHELL
BBC News Royal Correspondent